OXFORD MEDICAL PUBLICATIONS

The Challenge of Geriatric Medicine

The Challenge of Geriatric Medicine

BERNARD ISAACS
Emeritus Professor of Geriatric Medicine
University of Birmingham
and
Medical Adviser
JDC—Brookdale Institute of Gerontology
Jerusalem

Oxford New York Tokyo
OXFORD UNIVERSITY PRESS

Oxford University Press, Great Clarendon Street, Oxford OX2 6DP
Oxford New York
Athens Auckland Bangkok Bogota Bombay Buenos Aires
Calcutta Cape Town Dar es Salaam Delhi Florence Hong Kong
Istanbul Karachi Kuala Lumpur Madras Madrid Melbourne
Mexico City Nairobi Paris Singapore Taipei Tokyo Toronto
and associated companies in
Berlin Ibadan

Oxford is a trade mark of Oxford University Press

Published in the United States by
Oxford University Press Inc., New York

First published 1992
Reprinted 1997

A catalogue record for this book is available from the British Library

Library of Congress Cataloging in Publication Data
Isaacs, Bernard.
The challenge of geriatric medicine / Bernard Isaacs.
(Oxford medical publications)
Includes bibliographical references and index.
1. Geriatrics. I. Title. II. Series.
[DNLM: 1. Geriatrics. WT 100 173c]
RC952.I83 1992 618.97—dc20 91–28011
ISBN 0 19 262021 5 (pbk)

Printed in Great Britain by
Antony Rowe Ltd, Chippenham

Preface

Geriatrics began as the salvage of patients who had been confined to bed at home or in hospital, for months or years, without investigation, diagnosis, or rehabilitation. The clinical presentation of these patients was dominated by the 'Giants of Geriatrics': Immobility, Instability, Incontinence, and Intellectual Impairment. There was little guidance how to go about the work of salvage, so the early geriatricians developed their own systems. Even today, when research in geriatrics is flourishing, these early concepts continue to form the basis of the specialty.

The aim of modern geriatric medicine is to forestall the Giants of Geriatrics, by diagnosing and treating their causes. Unfortunately, the Giants still abound. So this book has been written to help doctors and nurses to deal with the Giants. It does not pretend to be a textbook, and needs to be supplemented by a textbook. But it tries to present in relatively simple terms ideas on how to approach and help patients who suffer from the Giants.

The book is selective, concentrating on areas of geriatric medicine which other books may not deal with. The criteria for omission are that the material is dealt with adequately in standard texts, or does not apply to elderly hospital patients, or is of local interest only, or is likely to change rapidly. This applies to demography, epidemiology, prevention, descriptions of diseases, therapeutics, policy issues, and social services.

To the chapters on the original four Giants, five others have been added, namely stroke, aphasia, visual impairment, auditory impairment, and depression. These are all written for the non-specialist. There are separate sections of notes and references at the end of each chapter. The main topics carry supplementary lists of aphorisms which may be helpful in teaching.

A friend who read the manuscript described the book as 'dogmatic and eccentric'. I was delighted: it was meant to be. My ideas have been influenced by many teachers and colleagues, but I would like to mention two who have been particularly helpful, Sir Ferguson Anderson and Alistair Main. And my wife Dorothy put up with a lot during the writing.

Birmingham and Jerusalem B.I.
March 1991

For Dorothy

Contents

1 *The Giants of Geriatrics*

The Giants of Geriatrics are:

- Immobility;
- Instability;
- Incontinence;
- Intellectual Impairment.

They have in common the qualities of:

- multiple causation;
- chronic course;
- deprivation of independence;
- no simple cure.

The Giants are disabilities, yet they are more than disabilities. They do not kill, but they diminish the value of living. They are the 'common final pathway' of the clusters of chronic diseases which frequently afflict very old people with diminished powers of recovery. They are called Giants because of the Gigantic number of old people whom they afflict and the Gigantic onslaught which they make on the independence of their victims.

The Giants make their victims dependent on others for care. So long as patients afflicted with the Giants remain in their own homes, the physically and emotionally demanding care which is required is provided by family, friends, neighbours and community services. Admission of the patient to a geriatric hospital or a nursing home may result from breakdown of care in the home; or it may be precipitated by an episode of acute illness, such as pneumonia, or a fall with a fracture of the femur.

The association between the presence of a disease and the occurrence of a Giant may be causal, contributory, consequential or coincidental; and one Giant may lead to another. For example, Parkinson's disease may cause Immobility and contribute to Instability. The patient may become incontinent as a consequence of the Immobility. Intellectual Impairment may be part of the process of Parkinson's disease, or it may be due to the coincidental presence of Alzheimer's disease.

Approaching the Giants

Doctors and nurses may experience difficulty in:

- taking the history;

- physical examination;
- diagnostic tests;
- defining objectives;
- providing treatment;
- making ethical decisions.

Taking the history is time consuming and often uninformative. Relatives have to be questioned, and previous notes have to be studied.

Physical examination reveals a wealth of signs, but it is not clear whether these contribute to the clinical picture or are incidental findings. Getting the patient out of bed to test balance and mobility takes time when no nurse is available to help.

Laboratory and other diagnostic tests. The more tests that are done the more abnormalities are discovered and the more difficult it is to interpret them. There is uncertainty about how far to go in pursuing the diagnosis when this may not lead to any difference in treatment or prognosis.

Defining objectives. It is easy to couch objectives in terms like 'To restore mobility' or 'To get the patient home'. This does not say how much mobility is to be restored; nor how the patient will manage at home.

Providing treatment. Drug treatment can be prescribed, but multiple pathology may mean multiple medication, which carries the dangers of poor compliance and multiple toxicity. Treatment involves not only drugs, but determining how the patient is to receive nursing care, physiotherapy, occupational therapy, speech therapy, family care, social services, psychological, and spiritual support.

Ethical issues. Many ethical questions have to be faced daily. Among these are:

- How far is it justified to investigate and treat patients who have multiple symptoms and a poor prognosis?
- Should a patient's wish to go home be respected when he or she is manifestly incapable of managing at home?
- How much strain is it justifiable to impose on caring relatives?
- When is a patient with Intellectual Impairment no longer able to make appropriate decisions about himself or herself, so that it becomes justifiable to deprive him or her of autonomy?
- How much weight should be put on the patient's request not to undergo rehabilitation, but to be left in peace?
- To what extent should the wishes of relatives be heeded, especially when they ask that nothing further should be done?
- What are justifiable risks in advising major surgery, e.g. amputation of a limb?
- When resources are limited what priority should be given for procedures such as cardiac surgery or renal dialysis in elderly people?

- Should tube feeding be embarked upon in patients who have poor life expectancy?
- When if ever should the order 'Do not resuscitate' be written into the patient's notes?

These questions raise profound issues of principle, but decisions have to be taken daily. It is possible to have clear and consistent rules, or do *ad hoc* decisions have to be made with the aim of doing the least harm?

Conclusion

There has been much discussion on whether geriatrics is just general internal medicine and nursing, or whether it is a specialty in its own right. The question will become superfluous when all doctors and nurses are as confident of managing the Giants of Geriatrics as they are of dealing with other diseases and disabilities which they encounter.

Notes and references

The term 'Giants of Geriatrics' was introduced by Isaacs (1975). Studies of these disabilities and estimates of their frequency are reviewed in the appropriate chapters of this book. The presentation of symptoms in elderly people is described in many textbooks (e.g. Besdine 1985; Kane *et al.* 1989).

The earlier literature on ethical issues in relation to the care of ill elderly people was reviewed by Cassel *et al.* (1986) under the following headings:

- foregoing life-sustaining therapy;
- decisions about resuscitation;
- nutritional support;
- impaired capacity to consent;
- autonomy and paternalism;
- research and informed consent;
- advance directives;
- justice—allocation of rsources;
- medical and social policy;
- quality of life;
- interdisciplinary issues;
- nursing homes.

Other issues which have attracted attention include euthanasia (Meier and Cassel 1983); the participation of families in decision-making (Sherlock and

Dingus 1984); and old-age abuse (English 1980; Quinn and Tomita 1986; Filinson and Ingman 1989).

Empirical research on ethical issues has been conducted on informed consent (Taub 1980); on the decision not to resuscitate (Schwartz and Reilly 1986); and on the attitudes of nurses and physicians to tube feeding (Watts *et al.* 1986) and to active intervention in terminal illness (Wolff *et al.* 1985). The clinical case against tube feeding in non-communicative, recumbent, terminally ill patients is presented by Campbell-Taylor and Fisher (1987).

The *British Medical Journal* published a series of articles (Gillon 1985) in which the ethical principles of beneficence, non-maleficence, autonomy, and justice are applied to the practical problems encountered daily by doctors. Elford (1987) presented a debate on ethical issues between practising geriatricians and professional ethicists.

References

Besdine, R. W. (1985). Clinical evaluation of the elderly patient. In *Principles of geriatric medicine and gerontology* (ed. W. R. Hazzard *et al.*) McGraw Hill, New York.

Campbell-Taylor, I. and Fisher, R. H. (1987). The clinical case against tube feeding in palliative care for the elderly. *Journal of the American Geriatrics Society*, **35**, 1100–4.

Cassel, C. K., Meier D. E., and Traines, M. L. (1986). Selected bibliography of recent articles in ethics and geriatrics. *Journal of the American Geriatrics Society*, **34**, 399–409.

Elford, R. J. (1987). *Medical ethics and elderly people.* Churchill Livingstone, Edinburgh.

English, R. W. (1980). *Elder abuse.* Franklin Research Center, Philadelphia.

Filinson, R. and Ingman, S. R. (1989). *Elder abuse; practice and policy.* Human Sciences Press, New York.

Gillon, R. (1985). An introduction to philosophical medical ethics: the Arthur case. *British Medical Journal*, **290**, 1117–19.

Isaacs, B. (1975). The Giants of Geriatrics. Inaugural lecture, University of Birmingham.

Kane, R. L., Ouslander, J. G., and Abrass, I. B. (1989). *Essentials of clinical geriatrics.* McGraw Hill, New York.

Meier, D. E. and Cassel, C. K. (1983). Euthanasia in old age: a case study and ethical analysis. *Journal of the American Geriatrics Society*, **31**, 294–8.

Quinn, M. J. and Tomita, S. K. (1986). *Elder abuse and neglect: causes, diagnosis and intervention strategies.* Springer, New York.

Schwartz, D. A. and Reilly, P. (1986). The choice not to be resuscitated. *Journal of the American Geriatrics Society*, **34**, 807–11.

Sherlock, R. and Dingus, C. M. (1984). Families and the gravely ill: roles, rules and rights. *Journal of the American Geriatrics Society*, **33**, 121–4.

Taub, H. A. (1980). Informed consent, memory and age. *Gerontologist*, **20**, 686–90.

Watts, D. T., Cassel, C. K., and Hickman, D. H. (1986). Nurses' and physicians' attitudes toward tube-feeding decisions in long term care. *Journal of the American Geriatrics Society*, **34**, 607–11.

Wolff, M. L. Smolens, S. and Ferrara, L. (1985). Treatment decisions in a skilled-nursing facility: discordance with nurses' preference. *Journal of the American Geriatrics Society*, **33**, 440–5.

Aphorisms

Of geriatrics and geriatricians

- Geriatric medicine is the treatment of underprivileged patients by underprivileged doctors in underprivileged buildings.
- Geriatrics is not so much a branch of medicine as a branch of archaeology. Just as the archaeologist recreates from a broken potsherd the magnificent vase of 2000 years ago, so the geriatrician sees in an incontinent, confused old woman the blushing young bride of 60 years ago.
- A parent wishes his child to be treated by an expert paediatrician. A child wishes his parent to be treated by an expert geriatrician.
- A geriatrician is a doctor with a soft heart, a hard head, a thick skin and a chip on the shoulder.
- With his soft heart he feels, with his hard head he decides, with his thick skin he fights, with his chip on the shoulder he suffers.

Of words

- It is better to change the fame of geriatrics than to change the name of geriatrics.
- The word 'geriatric' is not a noun and does not describe a person. Hospitals do not treat 'geriatrics' any more than schools teach 'paediatrics'.
- When we talk of 'the elderly' we generalize. When we talk of 'elderly people' we particularize.

Of old age

- Old age is having been born a long time ago. It is also having lived for a long time. These are not the same.
- Having been born a long time ago means bearing the features of a cohort. Having lived for a long time means sharing the experiences of a cohort.
- In old age the past is a rock, the present a shifting sand, the future a vaporous cloud.

- Old age is grief and anxiety and a struggle to pretend that it is neither.
- Old age is when you open your journal and look at the obituaries before the jobs; it is when retired people start to look young; it is when you eye all the beautiful girls on the beach in their bikinis and wonder which one you would like to have for a daughter-in-law.
- There are three stages of life after retirement, the Bronze Age, the Iron Age and the Stone Age. In the Bronze Age the body is fit and the mind is whole. The world and its sunshine are explored and enjoyed. In the Iron Age the body is fettered but the mind is free. The Zimmer is wielded but the spirit is strong. In the Stone Age both body and spirit are rock. The brain sleeps in the dark cave of the body, and there are no paintings on the wall.
- The person who says, 'Shoot me when I reach the age of 65' denies the existence of life after 65.

Of old people

- The characteristic feature of old people is diversity. There is no homogeneous biomass called 'the elderly'.
- Five-year-old children differ from one another more than new born babies do, because they have lived longer. Seventy-five-year-old people differ from one another even more, because they have lived even longer.
- Nice young people become nice old people. Nasty young people become nasty old people.
- Likeable old people attract care. Unlikeable old people repel care.
- Old people are like gold sovereigns. They bear the stamp and the values of the past.
- Old people may see themselves like old pennies, worn away, not worth much, but interesting.
- Most old people are young people in old bodies. Some young people are old people in young bodies.
- The lady of 93 who said that she was fed up with being marvellous meant that if it is marvellous to be normal, it must be normal to be abnormal.
- The old lady with pains all over her body said 'I shouldn't complain, there's thousands worse than me . . . but they are all dead.'
- The centenarian who was blind, deaf and bedfast said 'Do you want to know what it is to be a hundred? It's nothing.'
- The centenarian who sings to visitors caricatures old age.

Of treatment

- Old patients have old doctors. Old doctors use old drugs. Old drugs are toxic to old people.
- Set an objective and set a time to attain it.
- To know your patient visit him in his home, become his guest.
- Multiple pathology brings multiple opportunity.
- One straw breaks the camel's back; removal of one straw preserves the camel's back.
- The more a drug is used the more it is misused.
- Compliance means that the doctor knows best; non-compliance means that the patient knows best.

Of education in geriatric medicine

- The reason for teaching geriatrics is not the increase in numbers of old people but the increase in knowledge of old people.
- Observing old people in hospital is like teaching zoology in a zoo. Observing old people in their homes is like teaching zoology in the woods and fields.
- Those best qualified to teach geriatric medicine are patients, relatives, neighbours, and carers. Use them.
- Truth is the food of the mind: better a little digested than a lot regurgitated.
- He who has studiously watched an elderly person rise from a chair has supped the milk of geriatrics.
- Two students were asked what they had learned in geriatrics. One said 'It is not all doom and gloom.' The other remarked 'It's all out there.'

2 *Taking the history*

'The history' is the major source of information about the patient, exceeding what will be learned from physical examination and laboratory tests. Difficulties arise because of:

- the context;
- the nature of the patient;
- the nature of the illness;
- the preoccupations of the doctor.

The context

In the past, geriatricians were accustomed to meeting the patient for the first time in his or her own home, in the course of a pre-admission domiciliary assessment visit. This provided the ideal environment for getting to know the kind of person the patient was, the family among whom he or she lived, the physical features of the home, the presence of food, warmth, and cleanliness, the care taken of drugs, the quality of help obtained, the attitude of carers and the stress upon them. The discerning nose could identify the odours associated with urine or faeces, with sour milk and rancid butter, with cats and mice; or alternatively with furniture polish and disinfectant. The patient was host and not guest, and he or she was not intimidated or bewildered by being in hospital.

The pattern of work in the modern geriatric unit has reduced the frequency of pre-admission home assessment. When patients are admitted as emergencies, there may be little time for adequate history taking. The doctor learns to be economical in questioning, and to extract the maximum value from each question asked. The initial interview should be supplemented by a further one, once the patient has settled down in the ward.

The patient

Geriatric patients are sometimes described as 'poor historians' because they do not 'give a good history'. But an historian is a person who writes history, not one who provides the facts. The doctor working with elderly patients with indistinct memories is like the historian working with ancient documents with indistinct writing. Both have to use professional skill and discretion in collecting, analysing, evaluating, and selecting the information upon which the written 'history' will be based. It is difficult to obtain a concise, accurate, consistent, and complete account of the events which caused the patient's illness and led to admission to hospital. The doctor should think of him- or herself as 'interviewing the patient' rather than 'taking the history'.

The first consideration is: 'Does the patient understand me, and do I understand the patient?'

Testing comprehension

If the patient does not answer a question or gives an inappropriate answer the interviewer should consider:

- Does the patient hear what is said?
- If he (or she) hears does he understand?

If he does not understand is this because:

- He (or she) is not a native-English speaker.
- He has receptive dysphasia.
- He has brain failure.
- I am speaking too fast.
- I am using technical language.
- My questions are too long.
- My questions offer too many alternatives.
- I am not making my meaning clear.
- My questions are insufficiently specific.

If the patient does not reply is this because:

- He (or she) is not a native-English speaker.
- He has expressive dysphasia.
- He has anarthria.
- He is asleep or unconscious.
- He is over-sedated.
- He has brain failure.
- He is depressed.
- He feels he is being patronized.
- He does not wish to answer.

Testing credibility

The interviewer should not accept the patient's answers as necessarily representing the truth, the whole truth, and nothing but the truth. Answers may be inaccurate or incomplete because of impaired memory, comprehension, cognition, depression, and suppression.

Every interview should include a test of the patient's credibility. This should be done with even the most convincing patient; indeed the glib and articulate are those most in need of being tested. A diagnosis or assessment based on the testimony of an untested witness may be seriously misleading. Credibility is tested by asking the ordinary questions with which one would open a conversation with any stranger, such as the name, address, and age of the patient, the marital and family status, the numbers and names of children, grandchildren and great grandchildren, the names of daughters-in-law and sons-in-law, the former employment and the names of employers. These are open-ended questions which cannot be answered by 'Yes' or 'No'. If the patient has family photographs he or she should be asked to identify and talk about the people in them.

Interviewers should ask themselves:

- Did the patient answer the question?
- Did he (or she) evade the question?
- Did he refuse to answer?
- Did he ask someone else to help him?
- Was the answer complete?
- Was it consistent with other answers?
- Was it consistent with facts known to the interviewer?

If the patient is inaccurate or inconsistent or evasive or obstructive the interviewer should consider the following causes:

- The patient has a poor memory.
- He (or she) has poor power of concentration.
- He is fatigued or ill.
- He is unwilling to expose his impaired memory.
- He is trying to conceal his memory deficiency.
- He resents the question.
- He suspects that the interviewer is trying to make a fool of him.
- He cannot see the relevance of the question.
- He is angry.
- He is depressed.
- He wants to go to the toilet.
- He has brain failure.

If the patient fails the test of credibility, little reliance can be placed on his or her account of the illness. The facts must then be sought from relatives, friends, neighbours, the general practitioner, the home help, the community nurse or social worker.

Continuing the interview

If the patient passes the test of credibility, the interview is continued; but it should not usually be extended beyond about 15 minutes, as patients quickly become fatigued. Further information can be gathered on a later occasion.

Difficulties in obtaining the desired information may be due to:

- the expectations of the interviewer;
- the nature of the illness;
- the background of daily life;
- differing concepts of relevance;
- the quest for patterns;
- semantic poverty;
- excessive documentation.

The expectations of the interviewer

Some young interviewers feel uncomfortable in talking to old people. The gap in time and experience makes them uncertain what the patient is thinking. Interviewers may fear to raise topics like sex, incontinence, death, bereavement, and prognosis, which they fear may offend or upset the patient. But elderly patients are usually perfectly comfortable with these topics.

The interviewer should not expect too much from the patient. The 'history' of an 85 year old goes back a long time. It is tempting to select certain episodes and omit others, and thereby to present a coherent story which conforms to a pattern. This is particularly likely to happen when the patient or the doctor is thinking in terms of a process, or of a cause-and-effect relationship. An example of the first of these is when a gap of a year in a history of immobility is filled with the expression 'the condition steadily deteriorated'; whereas careful questioning may reveal a series of discrete episodes, with intervening periods of improvement. An example of the second is when a fall is attributed to bad weather, and the patient omits to mention that he or she had commenced taking an antihypertensive drug on the day before falling.

The nature of the illness

'Acute illness' in younger patients begins rapidly and prevents the patient from going to work. The Giants of Geriatrics appear slowly and insidiously in patients who have stopped going to work. Young patients can be expected to know exactly when the pain in the chest commenced; but elderly patients cannot say with any precision for how long they have been immobile. It is better to ask how the pattern of daily living has been altered by illness; and whether these changes were evident on memorable days like Christmas or their birthday.

The background of daily life

As an aid to establishing how the illness has developed the patient is asked:

Which of the following have you stopped doing because of your health:

- driving a car;
- travelling on a bus;
- crossing a busy street;
- visiting your children or relatives or friends;
- going to church or post office or shops or pub or club;
- working in the garden;
- climbing stairs;
- cooking;
- cleaning;
- making beds?

Why did you give them up?

Did you go to your doctor because you had to give them up?

Where did you spend last Christmas; or your last birthday? . . . and the one before?

How did you get there?

Relevance

Doctors and patients have different views of what is relevant. Patients may omit from the history vital events such as falls to which they attach no significance; or they may inflate the history with trivial events to which the doctor attaches no significance.

Patterns

Patients see patterns in their illnesses, interpret them as cause and effect, and present the story in conformity with their theory. They relate their symptoms to loss, separation, bereavement, trauma, weather, or medication. They may be right; but caution is needed to ensure that facts are presented as they occurred, and not in order to conform with the theory.

Doctors too may have their patterns, the 'textbook description of illness' with which they feel a sense of identity, and they may will the patient to yield a 'typical history'.

Semantic poverty

Many symptoms of disease in late life—for example postural instability, proprioceptive loss, sudden weakness, or transient loss of consciousness—have not been experienced before. Articulate patients describe these well; less articulate ones say that they were dizzy or that the arm is numb, or they collapsed or had a black-out. It is not always clear what is meant by these words, and the interviewer may misinterpret the patient's meaning. Asking 'What exactly do you mean?' is rarely fruitful. Clues may be found by watching the patient's gestures, or by establishing the circumstances in which the symptom appeared.

Documentation

Many patients in geriatric departments have masses of old notes. Doctors know that they should look through them, but this is tedious and time consuming. It is worth glancing at least at the first sentence on the first page of each admission for an indication of when and why the patient came in. This gives a broad picture of the course of the illnesses, and simplifies the search for clues to the current illness. No one can look through old notes without resolving that in future he or she will date accurately all comments, with the year as well as the day and the month.

Old notes may testify to the endurance of patients, who were recorded 10 years previously as immobile, doubly incontinent, confused and 'not for resuscitation'.

'Atrophy of the question mark'

The old notes often carry examples of 'atrophy of the question mark'. This occurs when, on a previous admission, the diagnosis was uncertain, and someone wrote down ?stroke or ?malignant, or ?multiple sclerosis. In the notes of the next admission the question mark has disappeared, and with it evidence of the previous uncertainty. The patient has acquired a firm diagnosis instead of an uncertain one. This simple phenomenon can cause misdiagnosis, mistaken prognosis, and mismanagement.

Common errors in history taking

Failure to elicit a 'good history' may be due to the following:

- inadequate assessment of the credibility of the patient;
- inadequate assessment of the patient's hearing;
- use of leading questions;
- failure to listen to answers;
- distortion of answers to fit preconceived ideas;
- 'Taking' the history rather than letting the patient 'give' it.
- working through a list of questions and paying insufficient attention to the answers;
- failure to follow up unexpected information;
- using language which the patient does not understand;
- asking questions that are too long or too complicated;
- asking questions that include the word 'or', e.g. 'Do you sleep right through the night *or* do you have to get up to go to the toilet?';
- using insufficiently specific language, e.g. 'Have you had problems with mobility?';
- using the continuous present tense rather than the perfect tense, e.g. 'Can you dress?' instead of 'Did you dress yourself this morning?';
- making unreasonable demands on memory, e.g. 'How long is it since you first felt dizzy?';
- accepting unquestioningly a distorted picture of reality;
- failure to question the improbable;
- failure to challenge overly optimistic presentations of current state;
- failure to identify denial responses;
- patronizing the patient by praising a correct answer to a simple question.

At the end of the interview the patient should be asked if he or she has any questions for the doctor.

Interviewing a relative

Relatives should be interviewed at the time of admission and during the patient's stay in hospital. The purpose of the initial interview is to:

- obtain or check the history of the patient's illness;
- learn the previous life-story of the patient;
- explore relationships within the family;
- understand how the life of the relatives has been and will be affected by the patient's illness;
- convey information to the relatives about diagnosis and prognosis;
- discuss the future management of the patient.

To obtain or check the history

In cases of brain failure the relatives are the only ones who know the whole story. The relatives of mentally normal patients can add to, confirm, or refute what the patient said.

To learn the previous life-story

The response to the illness is likely to be similar to the patient's response to previous challenges, illnesses, losses, or war experiences. This is particularly so in conditions such as stroke or fractured femur, where recovery very much depends on the patient's determination and effort.

To explore relationships

The relationship between patient and relatives is crucial to the understanding and management of patients afflicted by the Giants of Geriatrics. It is a professional privilege and a heavy responsibility to raise the curtain and view these relationships in action. They did not start with the patient's current illness but have developed over many years, and derive from fundamental family values.

The full exploration of these relationships requires patient work with the family by the social worker. However, the doctor and nurse obtain a sense of them in their dealings with the patient and the family. This helps them to understand the patient's response to illness; assists in discharge planning; and prevents hasty, superficial judgements about patients 'not wanting to get better' or families 'not wanting to know'.

To understand how the life of the relatives has been affected

Doctors and nurses have an understandable tendency to sympathize with the elderly patient and to see his or her interests as paramount. Patients are often on their best behaviour with the doctor, and present themselves as 'sweet old dears'. They do not always behave so to their own nearest and dearest. It is salutary to obtain the other side of the picture from relatives who have known the patient for very much longer than have the hospital staff. A few relatives may be manipulative, but the great majority are truthful and concerned. Those coming anew to the practice of geriatric medicine are often amazed to discover the great lengths to which families go in order to care for elderly relatives. Relatives welcome an opportunity of explaining to a doctor or nurse what the illness, or the personality of the patient, or the quality of their own lives are really like.

To convey information to the relatives

Relatives obtain information from many sources, including the patient, other patients, other visitors, staff on duty at the time of their visit, and the general practitioner. They may be bewildered by conflicting and inaccurate information. They may be unaware that the patient can walk or dress. They may believe that decisions have already been taken about his or her future. Doctors should ensure that relatives have an accurate picture of the situation, and should record in the notes what has been said, by whom, to whom and when.

To discuss future management

The Giants of Geriatrics seldom go away, and most patients need further care after leaving hospital. This is much easier to plan if the relatives are old friends of the hospital staff by the time discharge is in the offing.

Conclusion

Taking the history is the basis on which is built an understanding of the patient and his or her relatives; it is the means by which a relationship is forged between doctor, nurse and patient; and it is an education about human nature and the response to disability. It is the real intellectual challenge of medicine. The 'good historian' is the good physician.

Notes and references

Listening to the patient

Williams (1975) stressed the value of listening to what the patient does and does not say. Symptoms may be unrecognized because of their strangeness, neglected because of their banality, or ignored because they seem unimportant or irrelevant. The rational patient may suppress symptoms which appear to him or her to be irrational, through fear of being misunderstood. But 'neglected, unfelt, unrecognized and unspoken symptoms may be of the utmost importance in coming to a correct conclusion on aetiology'.

Self-perception

In long-term longitudinal studies there is a good correlation between a person's perception of his or her own health and subsequent mortality (Kaplan and Lamacho 1983). Mobility is a major component of self-perception of health (Jylha *et al.* 1986).

Medication

Compliance is impaird by the use of multiple medication (Hulka *et al.* 1975); and by impairment of vision, hearing, comprehension, and memory

(Stewart and Caranosos 1989). Compliance is in part attributable to 'intelligent non-compliance' by patients who experience adverse effects or lack of benefit from their medication. This in turn is related to the perceived benefit of treatment, the relationship between physician and patient, the level of explanation, and the nature and duration of the disease.

References

Hulka, B. S., Kupper, L. L., and Cassee, J. C. (1975). Medication use and misuse: physician–patient discrepancies. *Journal of Chronic Diseases*, **28**, 7–14.

Jylha, M., Leskinen, E., Alanen, E., Leskinen, A-L., and Heikinnen, E. (1986). Self related health and associated factors among men of different ages. *Journal of Gerontology*, **41**, 710–17.

Kaplan, G. A. & Lamacho, T. (1983). Perceived health and mortality: a nine year follow-up of the human population laboratory cohort. *American Journal of Epidemiology*, **117**, 292–304.

Stewart, R. B. and Caranasos, G. J. (1989). Medication compliance in the elderly. *Medical Clinics of North America*, **73**, 1551–63.

Williams, D. (1975). The borderland of epilepsy revisited. *Brain*, **98**, 1–12.

Aphorisms

- In history taking, the more questioning, the more obscuration; the more listening the more enlightenment.
- The historian is the person who asks the questions and writes the history, not the person who answers the questions and recounts the history.
- Patients are never 'bad historians'; doctors sometimes are.
- A good historian makes good use of bad material. A bad historian makes bad use of good material.
- The task of the historian is to take dusty old documents, blow away the dust, and decipher the ancient writing underneath. That is also the task of the geriatrician.
- The task of the historian is to sift through a mass of documents, reject what is immaterial and study what is material. That is also the task of the geriatrician.
- Much information is to be found in old notes.
- Remind yourself why the patient came in.
- Large, fat case-notes; small, thin patient.
- Beware the atrophy of the question mark. A patient with a brain tumour had a previous ?stroke. A patient with a spinal cord tumour had a previous ?multiple sclerosis. Between one admission and the next the question marks atrophied and the diagnoses were delayed.

- Atrophy of the question mark converts honest doubt to dangerous certainty.
- The suppression of doubt is the beginning of ignorance. The expression of doubt is the beginning of wisdom.
- Beware the old person who makes jokes. He is probably a humorist, possibly demented, and could be nasty.
- View the highly articulate with grave suspicion. They are probably pulling the wool over your eyes.
- Smoking and drinking are not part of the social history. They are social habits with medical consequences. They are part of the medical history.
- If a patient falls and breaks his femur, that is a medical problem. If a patient falls and does not break his femur, is that a social problem?
- Elderly patients are admitted to hospital not because of social problems, but because of medical problems with social consequences, or social problems with medical consequences.
- To understand Homer learn Greek. To understand old people learn their language.
- In lay language old people go off their feet, fall all over the place, wet themselves, become confused. In technical language old people suffer from immobility, instability, incontinence, and intellectual impairment.

3 *Physical examination and diagnosis*

Physical examination

To the routine, systematic examination which is conducted on those who suffer from the Giants of Geriatrics, special tests are added in relation to each of the Giants. These are described in the appropriate chapters.

The physical examination of the patient in bed is supplemented by the examination of posture, balance, and gait. If this is not convenient at the initial examination the procedure may be delayed until the patient is fit, but should not be omitted. Patients are liable to have incidental, contributory, but potential treatable disease in the eyes, ears, mouth, skin, and feet. The opportunity is taken of identifying and if possible correcting these.

Examination of the patient in bed

Being old means having been born a long time ago, and having lived for a long time. Having been born a long time ago means that the patient may have small stature, and may bear the marks of diseases which were prevalent and poorly treated in his or her childhood; such as rickets, kyphoscoliosis, congenital dislocation of the hip, poliomyelitis, osteomyelitis.

Having lived for a long time, the patient may have acquired war and industrial injuries which were not well treated by modern standards. An old stroke or arthritis may have been allowed to lead to deformities which nowadays would have been prevented; and outmoded operations may have been conducted, such as the visceropexies that were common in the 1930s, and the gastroenterostomies and gastrectomies of the 1940s and 1950s. A significant part of the medical history of an elderly person is written on his or her body.

The following are among the physical features which may be encountered in the very old.

The skin

The stereotype is that skin is thin, transparent, and purpuric, but surprisingly often it is like that of a young person.

Some common lesions

'Spots'. Among the profusion of flat or elevated spots, warts, and papillary protuberances which are often to be seen, it is necessary to distinguish the

common, innocent, seborrhoeic warts and fibromata from the rarer, premalignant and malignant conditions such as rodent ulcer, squamous epithelioma, acanthoma, lentigo, intra-epidermal epithelioma, and melanoma. Senile purpura should be distinguished from the rarer, genuine purpura and from traumatic bruising. The latter may indicate unreported falls, or may lead to the suspicion of non-accidental injury.

Intertrigo under the chin, breasts and abdominal folds is common in fat, sweaty women. Secondary infection with monilia (*Candida*) is especially to be sought in diabetics; it can be suspected from its smell and the rotting skin.

Tattoos give a colourful indication of former life-style. They are rarely a clue to previous hepatitis.

Pallor is common, and is probably associated with spending time indoors. It is a poor indicator of anaemia. The practice of calling pallor 'clinical anaemia' is to be deprecated. Anaemia can only be determined by blood examination.

Oedema. In the presence of gross oedema of the lower limbs the skin may blister or exude oedema fluid.

Broken skin from pressure, inflamed skin from chemical burns, and macerated skin from perspiration commonly accompany immobility and incontinence.

Pressure sores

The term 'pressure sore' embraces a range of destructive lesions of the skin and subcutaneous tissues of varied pathogenesis. These are also referred to as bedsores, decubitus ulcers, and brush burns.

Pressure sores tend to develop in parts of the body which doctors do not regularly inspect; and when they do their rounds the sores may have only just been dressed. Nurses have traditionally taken much of the responsibility for preventing, identifying, and treating sores. In departments of geriatric medicine, responsibility for the skin, as for all other parts of the body, is shared by all staff.

Lesions are described in terms of:

- site;
- depth;
- colour;
- base;
- edge;
- surrounds;
- pain.

Site. Common sites are the buttocks, natal cleft, and heels. Infarcted lesions caused by direct pressure are common over bony prominences in thin patients; while superficial brush burns may be found on the back or buttocks of heavy patients, in consequence of their having been pulled up the bed. In terminally ill patients, lesions can be found anywhere in the body.

Colour. The fixed, local erythema of the early stages proceeds to a dull or fiery reddening of the skin. Subsequently the skin may appear greyish or purple, raising the likelihood that there is extensive, deep infarction. This may have preceded the involvement of the skin. On the heel, purple discoloration may be followed without ulceration by the appearance of a black, gangrenous eschar.

Depth. Friction burns and tears of the skin may involve only the epidermis. Direct pressure sores commence with epidermal reddening, but may break through to involve dermis, fascia, and eventually bone.

Base. The tissue at the base of an ulcer may be:

- clean and pink, indicating active healing;
- covered with exuberant granulation tissue;
- yellow, due to necrotic fat or pus;
- black, due to necrotic muscle or tendon;
- exposed, with visible bone.

Edge. This may be:

- smooth and shelving, indicating healing;
- punched out and indolent;
- oedematous;
- overhanging, concealing extending necrosis under the edge.

Surrounds. The surrounding tissue may be swollen, discoloured, or 'boggy', suggesting widespread tissue infarction, even when the ulcer covers only a small area.

Pain. Most pressure sores are painful, especially those from brush burns and those on the heels. Patients who do not complain of pain may be obtunded by the systemic illness; or the sore may have developed in a denervated area, as in a diabetic with neuropathy.

Patients are most at risk of developing sores if they are very ill, have been immobile for a long time, and are very fat or very thin.

Other skin conditions

Varicose or gravitational ulcers. caused by venous insufficiency may have been present for many years, and have undergone many cycles of healing and breakdown. They are found in the lower third or half of the medial aspect of the leg; but very chronic ones may extend on to the back of the leg or become circumferential. If healing takes place the skin may become pigmented and paper thin.

Ischaemic ulcers. These tend to be small, painful, and distal. They are commonest in diabetics.

Scars. The body may show the residues of childhood scalds, adult shrapnel war wounds, late-life zoster, and lifelong surgery.

Livedo reticularis. Tartan mottling of the medial side of the legs and thighs is attributed to the habit of sitting hunched over the fire (*erythema ab igne*). The increased use of central heating in the homes of old people has made the condition less common.

Nails. Inspection of the finger- and toe-nails may reveal pitting due to psoriasis or rotting due to fungal infection. Dystrophy of the nails may be associated with poor nutrition, but is not caused by calcium deficiency. Finger-nails may be painted, or may show tobacco stains, dirt, or faeces. Toe-nails too may be painted. They may be greatly elongated and horn shaped (onychogryphosis); or thickened and heaped up. Both conditions indicate that they have not been cut regularly, evidence of combined physical and social impairment.

The hands

The *handshake* may be moist or dry, warm or cold, powerful or limp, steady or shaky, confident or fearful. It tells much about the patient.

Tremor should be sought at rest and during purposeful movement, noting its type, frequency, amplitude, and relation to intention. The common physiological type is of high frequency, is usually confined to the hands, is most evident during movement and when the patient is under stress, and may disappear on rest. It should not be confused with the tremor of Parkinson's disease, which is slower, slightly coarser, present at rest, and may diminish during movement.

Generalized wasting of *muscle*, especially of the back of the hand, is common in ill, old people and has no special significance. Wasting which is confined to palmar or thenar muscles suggests cord, root, or nerve compression or motor neurone disease.

Deformation of *joints* ranges from the appearance of painless nodes at the base of the distal phalanges, which have little effect on function, to the gross deformities and contractures found in rheumatoid arthritis.

Reddening of the thenar and hypothenar areas is not necessarily a sign of liver disease; nor does *transparency* of the skin over the back of the hand help in the diagnosis of osteoporosis.

Clubbing remains a significant clinical sign even in advanced old age. *Koilonychia* is on the contrary unrelated to iron deficiency anaemia.

The feet

The feet provide information on the effects of vascular insufficiency, neuropathy, arthropathy, trauma, bacterial and fungal infection, mechanical damage, and social neglect.

The feet should be inspected for the effects of mechanical damage, as an engineer might inspect the foundations of an old bridge, and marvel that it still carries traffic. Mechanical damage is seen on the skin as hardening and callus formation; on the tendons as shortening; and on the bones as displacement and disarray of the bony structures.

Shortening of the Achilles tendon may be related to the past wearing of high-heeled shoes. It causes painful pressure on the plantar aspect of the forefoot.

Lateral or valgus displacement of the toes is probably due to poor, lifetime patterns of gait established in childhood and not improved by wearing narrow shoes. Many old people show lateral displacement of all toes, not just of the great toe (hallux valgus). The second toe may be trapped under the third (piano toe). All toes may show hyperextension at the metatarsophalangeal joint and fixed flexion at interphalangeal joints (hammer toes). Pressure on the abnormal foot structure causes pain, callosity, and inflamed bursae. The patient's shoes should be inspected for evidence of pressure distribution during walking. Many have been stretched or cut to accommodate a metatarsal bursa.

The feet are inspected for evidence of infection and impaired circulation. The palpation of dorsalis and tibial pulses is not easy, and failure to detect them does not prove their absence. When the foot of a patient who is receiving diuretics becomes acutely inflamed the diagnosis of gout should be suspected.

The mouth

Ideally all patients should be seen by a dentist or oral hygienist, because of the high incidence of remediable oral pathology among elderly hospital patients. When this is not possible, doctors should diagnose and treat what they can, and refer selectively.

The teeth, gums, palate, tongue, pharynx, and mucous membranes are inspected for evidence of ulceration and bacterial and fungal infection, especially in smokers and those who complain of a painful mouth or a persistent bad taste in the mouth.

Dentures which cause pain or which do not fit or which do not provide an adequate bite should be reviewed by the hygienist or dentist. Dentures which move when the patient talks cause embarrassment and disparagement, and need modification or replacement. Dentures found in the locker or left at home may have been emotionally rejected, in which case oral rehabilitation and provision of new dentures may not succeed.

A smooth tongue devoid of papillae may be due to vitamin B deficiency, but is less significant than the same finding would be in younger patients. Ulceration and leukoplakia require investigation for possible malignant change. Discoloration of the tongue can look alarming but is often due merely to surreptitious sucking of sweets.

Angular stomatitis may be due to vitamin deficiency, but is more usually associated with ill-fitting dentures.

The neck

The range of active and passive movement should be examined in flexion, extension, lateral, and rotatory movements. Passive movement may generate impressive crackling and grating sounds, but their loudness does not correlate with the severity of the symptoms. Cervical spondylosis, in the absence of myelopathy and radiculopathy, has few physical signs. A sense of motion when the neck is rapidly turned may indicate positional vertigo; but patients find this manoeuvre unpleasant. If the movement is accompanied by short-lived nystagmus the suspicion of vestibular disease is raised.

The eyes

Outward and inward folding of the eyelids (ectropion and entropion) are easy to spot and to treat. Eyelashes may fall on to the eye causing pain, reddening, and tears. Excess tears may be due to blockage of the tear ducts. A dry, gritty eye may be associated with rheumatoid arthritis (Sjögren's syndrome).

Ophthalmoscopy is rendered difficult by small pupils and a roving eye. Cataract is easily spotted by shining the light of the ophthalmoscope obliquely on the pupil; and by noting opacity on direct examination.

Dilatation of the pupil may be required for retinal examination. The doctor should be familiar with the appearance of the optic cup in glaucoma, and with the signs of macular degeneration and retinal atrophy, as well as with the hypertensive and diabetic retinopathies.

The ears

These should be inspected for wax. Removal by syringing is time consuming and not always safe, pleasant, necessary, or effective. The removal of solid, impacted wax that is impairing the hearing is probably best left to the specialist.

The chest

The classical barrel chest of obstructive airways disease is still to be found, but the pigeon breast and Harrison's sulcus, consequences of childhood rickets, are now rarely seen.

The abdomen

The absence of a gap between the costal margin and the iliac crest is an indicator of kyphosis and spinal osteoporosis; and sometimes gives a clue to the cause of back pain.

Omphalolisthesis means the finding of a little stone in the umbilicus and suggests infrequency of ablution.

Examination of the *rectum* may reveal prostatic enlargement and rectal tumours, as well as a weak sphincter, a capacious rectum, and rectal contents which may vary in consistency from semifluid through pultaceous to the classic 'rocks'. Pultaceous faecal retention is associated with near intractable faecal incontinence.

The muscles

Muscle bulk is diminished. In obese patients, wasted muscle is replaced by fat, and the loss of muscle tissue is underestimated. In immobile patients, wasting is most evident in the vastus medialis muscle, which becomes concave. Fasciculation of muscle, especially in the upper limb girdle in men, raises the suspicion of motor neurone disease.

The back

A vertical, midline groove in the lumbar region is attributable to the habitual wearing of a tight corset. Loss of mobility of the back is common to many bony, arthritic, and soft tissue diseases of back structures.

Examination of the patient out of bed

Systematic observation of the patient's ability to rise from bed and chair, and to stand and walk, is part of the physical examination, and is of help in diagnosis and prognosis. Many patients are unfit to be tested at the time of admission. The test takes time and generally needs two people. If the doctor cannot perform the test, he or she should observe its performance by the nurse or physiotherapist.

The patient's performance is affected by environmental conditions. If lack of space, time, staff, or equipment cannot be avoided, allowance should be made for the resultant mechanical and psychological disadvantage. The test should be conducted in privacy, and the patient should be properly clothed and shod. Patients should not have to worry that they are being watched, and that their trousers might fall down or their slippers fall off.

The bed should be braked, so that it does not slide when the patient stands. The mattress should be firm so that the patient does not sink into it when he or she attempts to rise. If the bed is of variable height it should be adjusted so that patients' feet are firm and flat on the ground when they sit on the edge of the bed.

The chair should be placed at a right angle to the patient as he or she sits on the bed. It should have a straight back and a flat seat. The seat should not be so soft that patients sink into it when they attempt to rise; and it should not be raked back, so that patients' knees are higher than their hips, and their legs are flexed beyond a right angle. It should be of such a height that the patient has his or her feet firmly on the floor. The chair should have arms which patients can grip with their own arms flexed at about 135°.

The patient is then asked to:

- sit up in bed;
- swing the legs out of bed;
- sit on the edge of the bed;
- transfer into the chair;
- stand up from the chair;
- walk a few steps;
- turn round;
- return to the chair;
- sit down.

The doctor notes the speed and accuracy with which the test is performed, and the nature and severity of difficulties. Throughout the test the doctor stands unseen behind the patient, to protect them from a fall, but without distracting them.

Establishing the diagnosis

The Giants of Geriatrics are the outcome of multiple diseases in many organs. The traditional concept of 'making a diagnosis' is inappropriate. Instead all contributory medical, psychosocial, and environmental factors should be included in a problem list. For example, the problem list of a patient presenting with Immobility might include:

- ischaemic heart disease;

- chronic bronchitis;
- impaired nutrition;
- iron deficiency anaemia;
- osteomalacia;
- osteoarthritis of hips;
- hallux valgus;
- infected bunion;
- chronic glaucoma;

- ill-fitting shoes;
- stairs within house;
- stairs to outside from house;
- unsuitable walking-stick.

Diagnostic tests

The days when doctors were reluctant to do tests because of the patient's age are being replaced by the days when doctors are reluctant not to do tests despite the patient's age. Dr Richard Asher laid down the ground rules for laboratory tests when he said:

> Consider what you will do if the test is positive. Consider what you will do if the test is negative. If the two are the same, don't do the test.

Tests are justified if and only if they change management or prognosis. Avoidance of tests is justified if and only if they do not change management or prognosis. Age alone is not a consideration, because even at the age of 90 average life expectancy is several years, and that is a long time to suffer the effects of an inadequately diagnosed disease.

'Normal for age'

The interpretation of laboratory tests in geriatric medicine is bedevilled by the belief that what is abnormal in youth may be accepted as normal in old age. Common sense would suggest the opposite: if a finding is abnormal it cannot be normal, although it may be common.

There is an underlying practical difficulty. Most 'normal values' were established in populations of healthy young people. It is difficult to assemble large groups of healthy old and very old people for the determination of a normal range, and even more difficult to define 'healthy' among them. So results obtained in ill, old people cannot be compared with those from a normal population of the same age group. A finding from an ill, old person which is not in the normal range for young, healthy people might conceivably be in the normal range for people of his or her own age group if such figures were available.

Comfort range

An alternative is to define the circumstances in which the doctor feels comfortable. Taking the example of the haemoglobin level, the correct question to ask is not 'Is a haemoglobin of 11 g per cent in a man of 89 normal?' but 'Does the doctor feel comfortable with this level?' That is a subjective question, but geriatricians act subjectively whenever they balance the costs of investigation against the benefits of treatment. If the patient had a haemoglobin of 11.5 g per cent one year ago and is described as never having been a good eater the doctor is comfortable. If the patient had a haemoglobin of 14 g per cent three weeks ago when he was started on treatment with a non-steroidal anti-inflammatory drug the doctor is uncomfortable.

The same principle applies to blood pressure readings, physical signs, and the side-effects of drugs. It is unlikely that published 'normal' values have been derived from a population which was comparable to that of the patient. So doctors estimate the costs and benefits of intervention as best they can, and come to a considered decision that is right for their patient.

Conclusion

The examination and assessment of a patient who suffers from one or more of the Giants of Geriatrics uses the same methods and covers the same ground as the conventional medical examination; but pays more attention to areas of structure and function which are especially likely to yield findings of clinical significance in elderly patients.

Notes and references

Among reviews of the conditions discussed in this chapter are the description of skin conditions by Braverman and Fonferko (1982); of foot disturbances by Helfand (1978); and of the distinction between physiological and Parkinsonian tremor by Gresty and Findlay (1984) and by Griffiths (1985).

Presssure sores

The high prevalence of pressure sores in hospitals is documented by Barbenel *et al.* (1977); and the contribution of direct pressure, shear force, moisture, high temperature of the skin, sensory loss, and poor sitting posture are demonstrated by Maklebust (1987).

Many scales have been introduced for the assessment of the risk of developing pressure sores since the original Norton Score (Norton *et al.* 1962). These tend to overpredict, i.e. those with high scores do not necessarily develop sores (Gosnell 1987). There is no evidence that pressure sores are prevented by catheterization (Kennedy and Brocklehurst 1982).

Laboratory tests

Many laboratory test results from an ambulatory, elderly population lie outside the range of 'normal' derived from young people (Hodkinson 1977; Hale *et al.* 1983; Milne 1985); and results have to be interpreted with caution.

References

Barbenel, J. C., Jordan, M. M., Nicol, S. M., and Clark, M. (1977). Incidence of pressure sores in the Greater Glasgow Health Board Area. *Lancet,* **ii,** 548–50.

Braverman, J. C. and Fonferko, L. (1982). Studies in cutaneous aging. *Journal of Investigative Dermatology,* **78,** 434–45.

Gosnell, D. J. (1987). Assessment and evaluation of pressure sores. *Nursing Clinics of North America,* **22,** 399–416.

Gresty, M. A. and Findlay, L. J. (1984). Postural and resting tremors in Parkinson's disease. *Advances in neurology,* **40,** 361–4.

Griffiths, R. A. (1985). Neurological disorders of the elderly. In *Principles and practice of geriatric medicine* (ed. J. S. Pathy), pp. 627–94. John Wiley, Chichester.

Hale, W. E., Stewart, R. B., and Marks, R. G. (1983). Haematological and biochemical laboratory values in an ambulatory elderly population: an analysis of the effects of age, sex and drugs. *Age and Ageing,* **12,** 275–84.

Helfand, A. E. (1978). Foot health for the elderly patient. In *Clinical aspects of aging* (ed. W. Reichel), pp. 303–14. Williams and Wilkins, Baltimore.

Hodkinson, H. M. (1977). *Biochemical diagnosis of the elderly.* Chapman Hall, London.

Kennedy, A. P. and Brocklehurst, J. C. (1982). The nursing management of patients with long term indwelling catheters. *Journal of Advanced Nursing,* **7,** 411–17.

Maklebust, J. A. (1987). Pressure ulcers: etiology and prevention. *Nursing Clinics of North America,* **22,** 359–77.

Milne, J. S. (1985). *Clinical effects of ageing: a longitudinal study.* Croom Helm, Beckenham, Kent.

Norton, D., McLaren, R., and Exton-Smith, A. N. (1962). *Investigation of geriatric nursing problems in hospital.* National Corporation for the Care of Old People, London.

Aphorisms

- All abnormality is abnormal.
- An abnormal finding may or may not be significant but it is not normal.

- Abnormal findings do not become normal because other people have them; or because other old people have them.
- A handshake between doctor and patient is a courtesy and an investigation.
- A handshake reveals:
  the moisture of the febrile;
  the warmth of the vasodilated;
  the dryness of the hypothyroid;
  the coldness of cardiac failure;
  the power of the strong;
  the limpness of the weak;
  the inco-ordination of the cerebellar;
  the tremor of the parkinsonian;
  the apraxia of the hemiparetic;
  the disability of the Dupuytren;
  the pain of the arthritic;
  the tenderness of the gouty.
- If your hand hurts after the patient has shaken it he will be home in a month.
- Examination of horizontal patients by vertical doctors narrows the vision wondrously. Examination of vertical patients by horizontal doctors widens the vision wondrously.
- Say, 'Relax', and the patient tenses. Distract, and the patient relaxes.
- Tests are justified if and only if they change management or prognosis. Avoidance of tests is justified if and only if they do not change management or prognosis.

4 *Principles of rehabilitation*

This chapter deals with the principles of rehabilitation in a geriatric unit. The rehabilitation of patients suffering from the individual Giants is included in the appropriate chapters. Many patients who are admitted to departments of geriatric medicine or nursing homes regain lost function by their own efforts, or as a result of recovering from the disease which caused them to be disabled.

The rehabilitation of dependent old people is defined as 'the planned withdrawal of support'. Its aim is to place a competent patient in a congenial and prosthetic environment among confident carers.

Objectives

The outcome of rehabilitation is what the patient does and continues to do at home, and not merely what he or she can do in hospital. The objectives of rehabilitation reflect this view.

Objectives are immediate and ultimate. The ultimate objective is to return the patient to independent living at home. The immediate one is, for example, that by the end of the week he is able to put his trousers on without supervision.

Objectives should be specific, concrete, realistic, and dated, and should be known to patient, relatives, and staff. Attainment of an objective on time should be rewarded. Failure to attain an objective on time should be the subject of scrutiny. Was the staff's expectation faulty; or did some adverse factor intervene?

A patient made initial favourable progress after a stroke; but for two successive weeks the immediate objectives were not attained. A ward maid reported at the case conference that the patient had received a letter from a daughter in Canada telling her that she was undergoing a divorce.

Organization of rehabilitation

Staffing

In the ideal geriatric rehabilitation unit, physiotherapists, occupational therapists, and social workers are attached to the unit exclusively and for a prolonged period; and there is ready access to speech therapist, chiropodist, pharmacist, dentist, dietitian, prosthetist, and visual and audiological services.

There is help from a chaplain, unit clerk, volunteer organizer, clothing manager, and porters. All staff attend a weekly case conference, use a common documentation system, and are able to undertake home visits with the patient. The rehabilitation programme is organized by a unit manager, or by the nurse in charge, or by a member of the rehabilitation staff with the requisite authority.

Most rehabilitation of geriatric patients is undertaken in less favourable staffing circumstances.

The patient's day

The patient's day is timetabled so that he or she receives one or two hours of individual physiotherapy and occupational therapy each day, perhaps divided into two or three shorter sessions. This may be given in the ward, in a separate rehabilitation department, or in both.

The plan has to take account of the needs of patients and staff. Patients have to be helped out of bed and washed and bathed and fed and taken to the toilet and given drugs and drinks. They have to be seen by doctors and visitors. They may need blood tests, radiographs, and resting time. Staff may be dealing with emergencies or speaking on the telephone or talking to visitors or off sick or attending a study day or a case conference or on a home visit or engaged in an audit of their work; or they may have been moved to another ward. There is need for much flexibility and good will in working practices if the patient's needs are to be met.

Professional roles

The process of rehabilitation may bewilder patients. They may learn from the physiotherapist how to get in and out of bed independently, or from the occupational therapist how to put on their trousers. However, when the time to do so comes, they are in the presence of a busy night-nurse or a nervous relative, and may not have the opportunity of exercising their newly learned skills.

In assessing performance a distinction is drawn between 'peak' performance, which patients can achieve in optimal conditions, when there is no pressure of time and a skilled helper is present; and 'safe' performance, which they can achieve in more stressful circumstances. Patients, relatives, and day- and night-nurses should be asked to handle patients according to their best level of 'safe' performance.

Communication

Communication between hospital and community is along professional lines, i.e. doctor to doctor, nurse to nurse. Communication with patients and relatives may be insufficient or inconsistent. Patients may not know what is happening or what is wrong with them. Relatives may have different expectations of the outcome of hospital stay from one another, from the patient, and from the staff. Uncertainty and bewilderment may impair motivation. Units should develop clear working practices on who tells what to whom, particularly so that ample warning is given on when patients will be discharged, what will be their needs after discharge, and for how long dependency is likely to continue.

The weekly case conference offers an opportunity for communication among staff, and for planning communication with the patient, the relatives, and the community staff. The objectives are:

- to review and document progress;
- to measure the attainment of objectives;
- to reset objectives;
- to decide what is to be communicated to others and by whom;
- to review unit performance;
- to provide a forum for mutual education.

The doctor need not necessarily chair the meeting. Night staff should be present if possible. Relatives and community staff are encouraged to attend when an important decision is to be made. Sometimes patients attend to take part in the discussion on their own future. This ought to be the norm, but the atmosphere may be intimidating. The patient's views should be ascertained and presented at the meeting, and any proposals for the patient's future management which result from the meeting should be discussed with him or her.

Documentation

In addition to the records kept by each profession, there is a need for shared documentation which gives a rapid indication of individual patient progress, and provides instructions to staff on the degree of assistance or supervision required. This document can also be used as a management tool to assess the work-load on the ward. A standard assessment procedure such as the Barthel index may be chosen; or each department may devise a document for its own purpose.

For those who prefer to develop their own scheme the simplest system classes patients as independent, partially dependent, or dependent; and describes them by letters, numbers or colours: green (Go), yellow (Caution), and red (Stop). This system allows a rapid assessment of the state of patients in a ward but does not distinguish between different activities, does not specify how much help is needed by the partially and wholly dependent, and is not sensitive to improvement or deterioration within a grade. These difficulties can be dealt with by scaling separately different functions, such as mobility, transfers, continence, self-care, feeding, behaviour, mental state. Or the patient's ability to perform individual tasks, such as rising from a chair, dressing the upper part of the body, grooming, going to the toilet, may be represented on a scale which describes the amount of help needed.

The scores for separate functions can be added to give a single number, as is done in the Barthel scale. Assessments may be expressed on a flow chart, a graph, or by colour coding. The chart or graph can be displayed at or above the patient's bed, in the nursing station, or on a computer. The colour code can be charted, graphed, placed against the patient's name on the list of patients or on the door of his/her room or on the bed or bedside chart or wrist-band.

Ambitious charting procedures may collapse because they take too much time away from treating patients. Their successful implementation requires consideration of how much work is involved, who maintains the record, where it is kept, how frequently it is updated, and what use is made of it clinically and managerially.

Home visits

Home visits are undertaken usually by the physiotherapist and/or the occupational therapist close to the end of rehabilitation. Their purpose is to assess patients' performance in their own environment; to afford a glimpse into their past life and into their current surroundings; and to identify environmental and psychological barriers to mobility, which may not have been evident in hospital. The visit gives patients confidence in their ability to function again in their own home, and is educative for professional staff.

The therapist uses a check-list of daily living activities, which includes the following:

- access to the house;
- unlocking and locking the front door;
- access to the garden;
- unlocking and locking the back door;
- opening, closing, and locking windows;
- switching on and off lights;
- operating heating appliances;
- opening and closing cupboards and drawers;
- using cleaning equipment;
- using the toilet;
- using the washing hand-basin;
- using the bath or shower;
- making the bed;

- getting in and out of chairs and bed;
- using a kettle;
- using kitchen equipment;
- making hot drinks and snacks;
- filling a hot water-bottle;
- operating electrical equipment;
- dressing and undressing (usually assessed in hospital);
- reading a newspaper;
- switching on and off a radio or television set;
- taking part in a hobby.

The therapist observes and records the skill, competence, safety, and speed with which the patient conducts these activities. She also determines whether the carers have sufficient self-control and detachment to stand back and watch, or whether they are anxious and too ready to 'help'.

Recommendations include changes in the positioning of furniture, adaptations to furniture, and provision of aids, appliances, and minor adaptations like the fitting of hand-rails.

More radical changes may come under consideration before the return home of a patient who has suffered a major change in function, but who has a long life expectancy, e.g. a stroke victim. These include:

- installation of a chair-lift;
- building a downstairs bathroom, toilet, or shower;
- knocking down inside walls;
- removing or rehanging doors;
- reorganizing the kitchen;
- laying ramps.

These changes can be very successful; or they can be frustrating, expensive, disruptive, and painfully slow because of the need for planning permission, applications for grants, the discovery of unexpected defects in the building, and unsatisfactory workmanship.

Moving house may offer increased comfort, access, warmth and company; but should be recommended only when it is eagerly desired, when there is strong, continuing family support, when the physical advantages are outstanding, and when the gains outweigh the loss of a familiar, internal and external environment and the support of neighbours.

Attitudes to rehabilitation

Negative attitudes of patients

Patients are sometimes described as 'lacking motivation'. The staff expect the patient to be able to do more than he or she achieves. A few patients are lazy, and prefer someone else to care for them. This may be a lifelong habit; or it may stem from the feeling expressed as 'I have done things for others all my life; now it is time for others to do things for me.' But some patients who apparently lack motivation are physically or mentally ill, and should be reassessed for evidence of anaemia, heart failure, renal failure, malignant disease, or depression.

Attitudes of carers

Some relatives are inclined to over-protect, and exhortations to allow patients to do things for themselves are ignored. These carers expect to 'care', which they interpret as to 'help'. They must accept the value of the patient's independence, even if it diminishes the carer's role.

If the patient is confident but the relative is anxious, the patient becomes anxious. If the patient is anxious but the relative is confident, the patient becomes confident.

Conclusion

Rehabilitation requires professional skill in the diagnosis and treatment of physical and psychiatric illness; sensitivity to the needs of ill, old people and their families; and managerial skill in working with others in a well-planned and efficiently run unit.

Notes and references

Measures of function

Rating scales express a wide range of physical, mental, and behavioural functions in terms of numbers or grades. Their value and limitations were discussed with great wisdom by Townsend (1962). Israel *et al.* (1984) listed 245 published 'instruments' for geriatric assessment; and Feinstein *et al.* (1986) found 43 published measures of activities of daily living (ADL). These were reviewed and analysed by Kane and Kane (1981).

The Katz ADL scale (Katz *et al.* 1963) was among the first of these, and its simplicity has assured its continuing popularity (Katz and Stroud 1989). Lawton and Brody (1969) added the concept of 'instrumental activities of daily living', which includes the ability to use the equipment necessary to the running of a home.

Among other well-known scales are the Stockton geriatric rating scale (Meer and Baker 1966), the Clifton assessment procedure for the elderly (Pattie 1981), and the Older American Resources and Services (George and

Fillenbaum 1985). More closely related to physical rehabilitation is the Barthel index (Mahoney and Barthel 1965), which is recommended by Wade and Collin (1988) for general adoption. Simpler than any of these, and suitable for community and follow-up studies, is the grading of mobility into five stages: non-ambulant; limited or unlimited mobility in the house; limited or unlimited mobility in the community (Hoffer *et al.* 1973).

Guralnik *et al.* (1990) have reservations about tests in which the patient or a proxy describes what he or she can do; and prefer tests which require performance of an action under observation, such as the 'get-up and go' test (Mathias *et al.* 1986); measurements of walking speed and stride length (Pettmann *et al.* 1987; Wolfson *et al.* 1990); the ability to stand up from a chair (Friedman *et al.* 1988); and an index compiled from a number of simple clinical performance measures (Tinetti 1986). These tests are performed quickly, are objective, and correlate well with more complex measures of gait and balance, with one another, and with outcome.

Evaluation of rehabilitation

Controlled clinical trials in which one regimen of treatment is compared with another present methodological difficulties. Single-case studies, in which two regimens are compared in the same patient, partially overcome this difficulty (Edmans and Lincoln 1989; Wagenaar *et al.* 1990).

References

Edmans, J. A. and Lincoln, N. B. (1989). Treatment of visual perceptual deficits after stroke: four single case studies. *International Disability Studies*, **11**, 25–33.

Feinstein, A. R., Josephy, B. R., and Wells, C. K. (1986). Scientific and clinical problems and indexes of functional disability. *Annals of Internal medicine*, **105**, 413–20.

Friedman, P. J., Richmond, D. E., and Baskett, J. J. (1988). A prospective trial of gait speed as a measure of rehabilitation in the elderly. *Age and Ageing*, **17**, 217–35.

George, L. K. and Fillenbaum, G. G. (1985). OARS methodology: a decade of experience in geriatric assessment. *Journal of the American Geriatrics Society*, **33**, 607–15.

Guralnik, J. M., Branch, L. G., Cummings, S. R., and Curb, J. D. (1990). Physical performance measures in aging research. *Journal of Gerontology*, **46**, M141–46.

Hoffer, M. M., Feiwell, E., Perry, J., and Bennet, C. (1973). Functional ambulation in patients with melingomyelocoele. *Journal of Bone and Joint Surgery*, **55A**, 137–43.

Israel, L., Kozarevic, D., and Sartorius, N. (1984). *Source book of geriatric assessment*. Karger, Basel.

Kane, R. L. and Kane, R. I. (1981). *Assessing the elderly: a practical guide to management*. Lexington Books, Lexington, MA.

Katz, S. and Stroud, M. W. (1989). Functional assessment in geriatrics: a review of progress and directions. *Journal of the American Geriatrics Society*, **37**, 267–71.

Katz, S., Ford, A. B., and Moskowitz, R. W. (1963). Studies of illness in the aged: the index of ADL: a standardized measure of biological and psychosocial function. *Journal of the American Medical Association,* **185,** 914–19.

Lawton, M. P. and Brody, E.M. (1969). Assessment of elderly people: self maintaining and instrumental activities of daily living. *Gerontologist,* **9,** 179–86.

Mahoney, F. L. and Barthel, D. W. (1965). Functional evaluation: the Barthel index. *Maryland State Medical Journal,* **14,** 61–5.

Mathias, S., Nayak, U. S. L., and Isaacs, B. (1986). Balance in elderly patients: 'get-up and go' test. *Archives of Physical Medicine and Rehabilitation,* **67,** 387–9.

Meer, B. and Baker, J. (1966). The Stockton geriatric rating scale. *Journal of Gerontology,* **21,** 372–403.

Pattie, A. H. (1981). A survey version of the Clifton Assessment Procedure for the Elderly. *British Journal of Clinical Psychology,* **20,** 173–8.

Pettmann, M. A., Linder, M. T., and Sepic, S. B. (1987). Relationships among walking performance, walking stability and functional assessment of the hemiplegic patient. *American Journal of Physical Medicine,* **66,** 77–90.

Tinetti, E. M. (1986). Performance oriented assessment of mobility problems in the elderly patients. *Journal of the American Geriatrics Society,* **34,** 845–50.

Townsend, P. (1962). *The last refuge.* Routledge and Kegan Paul, London.

Wade, D. T. and Collin, C. (1988). The Barthel index: a standard measure of physical disability? *International Disability Studies,* **10,** 64–7.

Wagenaar, R. C. *et al.* (1990). The functional recovery of stroke: a comparison between neurodevelopmental treatment and the Brunnstrom method. *Scandinavian Journal of Rehabilitation Medicine,* **22,** 1–8.

Wolfson, L., Whipple, R., Amerman, P., and Tobin, J. N. (1990). Gait assessment in the elderly: a gait abnormality rating scale and its relation to falls. *Journal of Gerontology,* **45,** M12–19.

Aphorisms

Of rehabilitation

- Rehabilitation is the planned withdrawal of support.
- The motto of rehabilitation is 'hands off the patient'.
- Rehabilitation, like surgery, needs a no touch technique. Hands on the patient, like bacteria in the wound, are infectious and deadly.
- Unplanned rehabilitation is like an unplanned holiday. You don't know where you are going until you get there, and when you get there you don't know if that is where you want to be.
- Sudden death from overactivity is much feared and rarely seen. Gradual death from underactivity is little feared and much seen.

Of motivation

- Motivation is what patients are said to lack when they fail to get better.

- Lack of motivation to get better may be due to lack of a motive to get better.
- Lack of motivation in the patient may mean lack of imagination in the doctor.
- Patients lack motivation to do what the rehabilitation team wants; the rehabilitation team lacks motivation to do what the patient wants.

Of objectives

- Distinguish between objectives and objections. The patient's objection to the staff's objective is the patient's assertion of his or her own objective.
- When the staff objects to the patient's objective and the patient objects to the staff's objective, that is the time for re-examination.
- Planning someone else's discharge is like doing their shopping. You need to know what they want, and what price they are prepared to pay.

Of feet, shoes, and trousers

- Tower blocks cannot be built on bad concrete. Rehabilitation cannot be built on bad feet.
- A good rehabilitation department is one where the relatives bring in the shoes.
- Slippers make relatives feel kind; shoes make them feel unkind.
- Teaching patients to walk in ill-fitting slippers is like teaching actresses to smile in ill-fitting dentures.
- Patients do not walk at their best when they fear that their trousers will fall down.
- It is easier to get a liver scan in a hospital than a pair of braces.

Of prognosis

- You can make a prognosis by glancing at patients sitting. The upright will stand and walk; the sloucher will stand and fall; the curled are asleep; the head in hands are depressed; and the diagonals have had a stroke.
- Beware a handkerchief in the hand, a handbag on the arm, food in the mouth, teeth in the jar, the hearing aid in the drawer, and the artificial limb under the bed; all betoken a poor prognosis.

Of chairs

- Standard chairs are ideal for standard people; but for others they are too big, too small, too high, too low, too broad, too narrow, too soft, or too hard.
- For the young a chair is something to get into; for the old a chair is something to get out of.
- An easy-chair is hard to get out of; a hard chair is easy to get out of.
- Old people should have their feet measured so that their shoes fit, and their bottoms measured so that their chairs fit.

5 *Immobility 1: definitions and history*

The Giant Immobility deprives elderly people of the enjoyment of living by limiting life-space.

Definitions

Life-space

Life-space is where a person spends his or her days and nights. It is potentially boundless, but is limited by time, money, and health.

Life-space zones

The life-space of disabled people can be conceived of as a series of concentric zones, namely:

- bed;
- bedroom;
- bathroom;
- the rest of the house;
- the garden or courtyard;
- the block in which the house is situated;
- everything that is on the other side of a traffic-bearing street.

The outermost zone includes shops, the post office, the homes of family and friends, pubs, churches, cinemas, parks, buses, trains, airports, the rest of the world.

Movement from an inner to an outer zone requires passage through 'gates'. These are:

- getting out of bed;
- rising from a chair;
- walking;
- climbing stairs;
- crossing a street;
- using a vehicle.

These activities require 'programmed' movements. The ability to perform these activities, or open these gates, determines the extent of the life-space.

Mobility

The word 'mobility' is used here to mean the ability to move independently through life-space. This definition includes the use of a walking aid held by the subject, but excludes the use of a wheelchair or a motor car.

Immobility

Immobility is impairment of the ability to move independently, which results in limitation of life-space.

Programmed movement

Programmed movements are executed by a rich sequence of contraction and relaxation of countless muscles. Correct performance demands that each participating muscle and joint respond exactly as and when its contribution is required. A central, controlling mechanism is kept informed of the progress of the sequence. In the event of any default, instant substitution or correction is required, or the whole sequence may be halted.

Failure of the programme of gait can cause the subject to fall. Failure of the programme for rising from a chair can cause the subject to become 'stuck' half-way or to fall back into the chair. Programmed movements are controlled by:

- a master command, which instructs the subject to 'rise' or to 'walk';
- a 'start' command, which takes into account the bodily and environmental conditions at the start of the movement;
- 'modifying' commands, which alter the progression of the programme in the light of events that occur during its execution.

In gait the master command activates the 'walk' sequence; the 'start' command takes into account bodily functions such as muscle strength and joint pain, and environmental features such as ground surface, footwear, and lighting; the 'modifier' speeds or slows the motion, and initiates changes in direction or in height of foot swing to avoid encounters with other pedestrians, puddles, and irregularities of the surface.

In rising from a chair the master command instructs the subject to rise; the start command takes account of the height of the chair and the subject's strength, in order to ensure that sufficient mechanical force is generated to project the subject from the chair; the modifier switches the programme to deal with an unexpected change, such as the chair or the feet slipping.

The successful completion of programmed movements is jeopardized by:

- environmental conditions that impose too great a mechanical demand, such as sloping ground or too low a chair;

- bodily conditions that impair the execution of the programme, such as painful joints or weak muscles;
- central impairment of the initiation, execution, supervision, or modification of the movement, as a result of disease in the brain-stem, cerebellum, basal ganglia, or elsewhere in the motor or proprioceptive mechanism.

What the doctor needs to know

The history and physical examination seek to establish the severity, duration, mechanism, and cause of the patient's immobility. The severity of immobility is established by determining what zones of life-space the patient occupies, its duration by the extent of the life-space in the remote and recent past, and its mechanism by which 'gates' the patient can open.

Causes of immobility

Mobility may be restricted temporarily, intermittently, or permanently. It may be restricted more by night than by day, and more in winter than in summer. The causes of immobility include physical, psychological, and environmental factors, as follows.

Physical symptoms

- Breathlessness
- Fatigue
- Pain in hips, knees, back, feet
- Angina
- Claudication
- Vertigo
- Instability
- Visual impairment
- Auditory impairment

Psychological factors

- Depression
- Anxiety
- Agoraphobia
- Forgetfulness
- Fear of incontinence
- Fear of falling
- Fear of being conspicuous
- Fear of attack on person
- Fear of attack on property

- Nowhere to go
- No one to go with

Environmental factors
- Ill-fitting shoes
- Stairs
- Hills
- Traffic
- Broken ground
- Adverse weather

Prominent among the many diseases associated with immobility are brain failure, arthritis, stroke, fracture of the lower limb and pelvis, osteoporosis, amputation, parkinsonism, cardiac failure, obstructive airways disease, and foot disorders.

History taking

What the doctor needs to know
- What is the present life-space?
- What was the past life-space?
- If there has been a diminution, how did this occur, over what time course, in what stages?
- Why did it occur?
- What were the physical, psychological, and environmental causes?
- What are the physical, psychological, and social consequences?
- What has been done abou the restriction and its consequences?

Loss of life-space

The questioning begins by mapping the patients' world, identifying where they live, in what kind of house, how they reach the outside world, where places of importance are situated.

The patient is asked where they went and what they did, rather than where they go and what they do. The question 'Do you go to the shops?' may be answered by 'Yes'; but this might mean that the patient went to the shops in the past but does not do so now. If the question, 'Did you go to the shops yesterday?' is answered by 'Yes', this is unambiguous.

Concrete questions yield concrete answers. The patient is asked to which shop s/he went, where the shop was, how s/he reached it, what s/he brought back, what difficulty s/he had. If s/he went only to the bathroom s/he is asked how often, with what help, and with what difficulty.

The questioning assists the patient's memory by relating past events and functions to significant times in the patient's life. For example, s/he is asked where and how s/he spent his/her last birthday or Christmas; how s/he got there; whether s/he walked, went by car, taxi, or bus; how s/he used the toilet; whether s/he managed stairs; what help was received. This is preferable to questions of the type 'When did you last...?' or 'How long is it since...?'

Patients may be able to attach a date to the last time they undertook a favourite activity, such as playing golf or bowls, going for a walk in the country or park, visiting family and friends, going to the bank, post office, hairdresser, pub, shops, church, club, or library.

Not going out at night or in bad weather or in areas notorious for crime may have no medical significance, since many well elderly people prefer to stay at home in these circumstances. On the other hand these are sometimes advanced as reasons for not going out by people who have significant medical or psychological reasons for remaining at home.

Causes of restricted life-space

There is no direct relation between physical impairment and restriction of life-space. A person who has nowhere to go may go nowhere without being physically impaired; while a person with somewhere to go may go somewhere, despite being physically impaired. One old man endures severe symptoms to collect his pension; another uses mild symptoms as an excuse for not doing so.

Psychosocial factors which contribute to immobility include, in addition to those listed above, the fear of being assaulted or insulted; of needing to go to the toilet while out and not being able to find one; of falling and spilling the contents of the shopping bag; of having to stop and become conspicuous because of an attack of dyspnoea, angina, or claudication; as well as fear of traffic and of the effects on health of bad weather.

Within the house, people may cease going upstairs to bed because of fear of falls, fires, and burglars; and because they do not want to sleep alone in the room that was once shared with a deceased partner.

Consequences of restricted life-space

Physical consequences

The physical consequences of restricted life-space include:

- 'deconditioning', a general sense of slowing down, feeling tired and unfit;

- sleeping a lot during the day;
- tending to stumble and fall;
- eating too much;
- drinking too little, because of difficulty in reaching sources of fluid, and because of a desire to limit the number of journeys to the toilet;
- smoking too much, because there is nothing else to do;
- gaining weight, because of fluid retention, increased intake, and reduced activity;
- dependent oedema;
- excoriation of skin;
- pressure sores;
- incontinence, because of delay in reaching the toilet.

Psychological consequences

The psychological consequences of restricted life-space include resentment, frustration, irritability, grief, and depression.

Immobile patients may read, write, knit, sew, paint, compute, collect for charities, conduct prayer meetings, watch television, complain that 'I sit here looking at these four walls'; or exist uncomplainingly in languid inactivity.

The sense of loss is keenest in those who valued mobility most highly, loved outdoor hobbies, driving, and helping others. Those who rarely stirred from their television sets may be content not to have to. Those who live alone may be determined to remain independent; while those who live with others may be more ready to accept help.

Sleeping in the chair

Some old people with impaired mobility spend the night sleeping in a chair. Their reasons include warmth, economy in heating, fear of fire, fear of burglary, difficulty in ascending and descending stairs, reluctance or inability to remove clothing, and difficulty in getting in and out of bed. Sometimes the reason is emotional, their having abandoned the marital bed on the death of the spouse.

Sleeping in a chair further limits mobility by disuse; and frequently results in gross leg oedema and ulceration. The presence of these signs should raise the suspicion that the patient never goes to bed, which is rarely mentioned spontaneously. The oedema can be relieved by conventional treatment in hospital, but recurs if the patient persists in the habit of sleeping in the chair at home.

Social consequences

The social consequences of restricted life-space include the use of help in daily living; and the effect of disability on family members.

Daily-living needs include:

- *domestic*—cleaning, shopping, cooking;
- *personal*—dressing, bed-making and being taken to the toilet;
- *intimate*—washing, assisting with toileting and cleansing excreta.

The intervals between periods of need for help are classed as:

- *long*—exceeding one day;
- *short*—less than one day;
- *critical*—very short or unpredictable.

The type of need and the interval between necessary periods of help reflect the degree of immobility. Patients with mild immobility may have no needs, or may require long-interval domestic services only. Those with moderate impairment of mobility require long-interval services to assist in domestic tasks, but may not need other services. Patients with severe immobility require short-interval or critical-interval services to meet domestic, personal, and possibly intimate needs.

What help has been received?

The patient is asked about drugs, surgery, physiotherapy, chiropody, prostheses, and environmental changes.

Prostheses include walking-sticks ('canes'); three- or four-legged or wheeled walking-frames ('walkers'); and wheelchairs, which comprise self-propelled, 'transit', and electrically operated, indoor or outdoor varieties. Not all prostheses are appropriate, and patients should be asked:

- When and by whom they were recommended.
- Whether they were prescribed, purchased, borrowed or inherited.
- How much they are used.
- Whether they help.

Changes in the environment include:

- change of house to a simpler apartment or to the home of a relative;
- a move of bedroom;
- moving the bed into a downstairs room;
- purchasing a new bed or chair;
- installing a downstairs bathroom, toilet, or shower;
- installation of a stair-lift;
- fitting hand-rails in the bathroom.

Physical examination

In addition to the examination in bed for the signs of cardiorespiratory, musculoskeletal, and neurological disorders, and for the presence of local or diffuse muscle wasting, the fully clothed and shod patient is tested for the ability to move from one life-space zone to the next, i.e. to rise from bed or chair, transfer, walk, and climb stairs.

Performance may be impaired by local, diffuse, central or environmental factors, as follows.

Local

- Pain or limitation of movement at joints, especially the large joints of the lower limbs
- Painful feet
- Fracture or dislocation
- Ulceration
- Ischaemia
- Neuropathy or myopathy
- Paralysis or paresis of limbs

Diffuse

- Muscle wasting due to malnutrition or systematic disease
- Weakness due to cardiac failure, anaemia, cachexia, malignancy, 'de-conditioning'
- Dyspnoea due to cardiac or respiratory disorder

Central

- Parkinson's disease
- Other movement disorders
- Apraxia, i.e. a failure to organize the movements in their appropriate sequence
- Impaired vision
- Impaired balance
- The late stages of dementia

Environmental

- The firmness of the mattress
- Its height above the floor
- The presence of bedside furniture
- The floor and foot coverings

Transfers

The patient is observed getting out of bed, transferring into a chair, from one chair to another, and standing up from a chair, as described in Chapter 3. Ideally the mattress should be firm and of the correct height; and the chair should be of the same height as the bed. When the start conditions are less than ideal, due account should be taken of the difficulty which this causes.

In health, transfers are achieved by a single, coordinated and uninterrupted sequence of movements. Patients with 'peripheral' lesions, e.g. a leg in plaster, or an artificial limb or a stiff hip, may achieve normal transfers by adapting the programme to the abnormal conditions. If, however, the mechanical demands of the transfer exceed the patient's power to generate and sustain the full programme, the sequence is broken, and the patient may fall back to the starting position. This may occur when the start conditions are mechanically disadvantageous.

Patients with 'central' lesions, i.e. disease affecting the brain-stem, cerebellum, basal ganglia, or motor cortex, may be unable to generate the necessary sequence of movements. They make hesitant or inadequate attempts to execute a part of the programme, but become 'stuck' and abandon the attempt.

Walking

Observation of the gait gives insight into diagnosis, functional assessment, and prognosis. The quality of the gait is predictive of discharge potential and of survival. In observing the gait, attention is paid to:

- posture;
- speed;
- step length;
- asymmetry of step length;
- duration of double support;
- ground contact;
- stride width;
- out toeing;
- pelvic tilt;
- arm swing.

The features of the gait which are seen in the 'deconditioned' patient include:

- slowing;
- shortening of step length;
- lengthening of double-support time;
- lowering of the trajectory of swing;
- loss of heel strike;
- reduction of arm swing.

The factors likely to be responsible for abnormal gait include:

- weakness and wasting of glutei, hamstrings, quadriceps, leg extensors;
- spasticity or flaccidity in any of these muscle groups;
- pain at hip, knee, ankle, foot;
- stiffness at any of these joints;
- reduced range of movement at any of these joints;
- contracture at any of these joints;
- impaired muscle coordination;
- drop foot;
- amputation;
- inequality of leg length;
- ill-fitting shoes.

'Characteristic' gait patterns

The 'classical' patterns of gait abnormality in balance impairment, apraxia of gait, Parkinson's disease, arthritis, stroke, peripheral neuropathy, and other conditions are not always seen in elderly people with multiple diseases; and it is not easy to analyse why patients walk as they do.

Impaired balance

Patients with mildly impaired balance shorten their step length, slow their speed, widen their stride, and turn their toes outwards. Those with more severe impairment have an inconsistent 'irregularly irregular' gait, with a mixture of normal and abnormal steps. The patient may break into a run, or the feet may 'stick to the floor'. There may be a rapid change of speed or direction.

The stepping pattern can be restored to normal by supporting the patient and increasing speed. A convenient way of doing this is to allow the patient to hold on to a wheeled device (the case-note trolley serves the purpose), which is then pulled forward rapidly by an assistant.

Apraxia of gait

Patients with very irregular gait in whom central control appears to have broken down are sometimes said to suffer from 'apraxia of gait'. It is difficult to distinguish clinically between this condition and the abnormal gait of people with severe balance disturbance, and there is some doubt about whether 'apraxia of gait' is a specific condition. The subject is discussed further in Chapters 6 and 7.

Parkinson's disease

The classical 'festinant' gait of Parkinson's disease is not often seen in treated patients. While some parkinsonian patients 'chase the centre of gravity', it is more common to observe a tendency to fall back, because of inability to bring the centre of mass of the body sufficiently far forward.

Patients may suffer unpredictable but temporary arrest of progression during walking ('freezing'); as well as longer periods of total immobility (the 'off' periods of the 'on–off syndrome'). 'Freezing' may arrest the patient in a stable position, so that he or she stands still like a sentry; or the momentum may cause the trunk to continue its forward progression while the feet remain rooted to the ground, resulting in a fall on to the face and two black eyes.

Arrest of gait most often occurs when the patient approaches a door or other obstacle, at a time when the control of gait appears to change, as it were, from 'automatic' to 'manual'. Patients walk better when the environment is open and uncluttered.

Arthritis

The gait of the arthritic is determined by restriction of range of movement at affected joints; and by the desire to minimize pain. The pattern is individual and consistent, so that gait is 'regularly irregular'. Unloading a painful joint is achieved at the cost of overloading a painless one. The 'limp' should be analysed to see why the patient has modified the pattern, and to determine how best to correct the inappropriate loading. A waddling gait is seen in patients with arthritis of both hips; while those with unilateral hip disease tilt the pelvis down to the unaffected side and swing the affected thigh as they walk.

Stroke

The characteristic 'hemiplegic gait', with circumduction of the affected lower limb, is characteristic of unsuccessful rehabilitation of stroke. A well-rehabilitated stroke patient walks almost normally. This is described in Chapter 12.

Peripheral neuropathy

When this reaches the stage of distal muscle weakness and proprioceptive impairment, patients pick up their legs and stamp them down again in a characteristic 'prancing' fashion. Less severe cases with mild drop foot are suspected by observing the toes scraping the ground. Examination of the soles of the shoes indicates where excessive pressure has been applied.

6 *Immobility 2: rehabilitation*

Rehabilitation of the immobile patient begins with reconditioning. This is followed by training in transfers, walking indoors and outdoors, and climbing stairs.

Reconditioning is achieved by good food, adequate fluids, bowel regulation, pain control, sufficient sleep, avoidance of sedation, and attention to appearance, clothes, shoes, teeth, spectacles, hearing aids, hair, and shaving; and by exercises and resisted movements which tone up and strengthen muscles and which stimulate and accelerate slowed reactions.

Training and technique

Transfers

Patients are taught to transfer from bed to chair, from one chair to another, and from sitting to standing.

Bed to chair

This transfer can be made easier by reducing the task to subroutines. The sequence is:

- sit up from the lying position;
- remain seated with legs extended;
- swing legs round;
- lower legs to the floor;
- grasp the chair;
- swing on to the chair.

Chair to chair

For chair-to-chair transfers the two chairs are placed at right angles to one another, allowing the patient to stand, turn, and sit without having to take steps, and with minimal support. The patient wears shoes with roughened rubber or plastic soles which fit well and which provide good ground contact.

Standing

Patients may fail to release the hands from support before the body has reached the upright position. This is corrected by raising the height of the seat and by strengthening the muscles.

Walking

Before walking training is commenced, patients should be toileted, dressed, and shod. There should be a suitable, well-lit area of floor, free from distracting furniture and other people.

The patient who clutches or is clutched will not regain confidence. Unsteadiness is controlled not by gripping the patient's arm but by lightly touching the back of the head or the pelvis, as is described more fully in Chapter 8.

Walking aids

Walking aids are used to diminish the load on painful joints. Their use to assist balance should be minimal. Walking aids resemble drugs in being effective in the short term, toxic in the medium term, and addictive in the long term. They may cause abnormal posture or gait, and patients may become unnecessarily dependent on them. They should be prescribed for specific indications and for a limited period.

Stairs

Training in ascending and descending stairs is best done in the patient's home, where an assessment can be made of the need for hand-rails, and the state of the stair carpet can be inspected. Most patients who can walk can climb, and problems occur with the impulsive and the reckless rather than the handicapped. Good levels of lighting are essential, especially near the top and bottom, where the patient may misjudge the point at which the staircase ends and step into the air. Another visual hazard occurs at the point during descent when the view of the upper floor suddenly gives way to that of the staircase leading down to the lower storey. Falls may occur during descent at this point.

Walking-sticks can be used on the staircase. Patients who use walking-frames should keep one downstairs and one upstairs, and should use two hand-rails during ascent and descent.

When stair climbing becomes exhausting and hazardous some patients crawl up on their hands and knees and 'bump' down on their bottoms, preferring this to bringing the bed downstairs.

Going out in a car

Car journeys increase life-space. Training relatives in helping immobile patients into and out of a car is part of rehabilitation. In two-door cars the passenger door opens widely enough to enable those with arthritic hips to sit and swing the legs into the car. In four-door cars it may be easier for the patient to sit on the driver's seat, then lie down across the front seat and wriggle backwards until the legs are clear of the gear-stick and the sitting position can be regained on the passenger's seat.

Thought has to be given to where the car will be parked on arrival and departure, how much clearance there is for the doors, what help there is at each end, how the buildings and the toilets will be accessed, whether a wheelchair is required, where it will be stored, and what cover there is if the weather is wet.

Rehabilitation in diseases associated with immobility

Arthritis

Arthritis causes pain, stiffness, weakness, muscle spasm, and limitation of movement. Functional impairment of one joint overloads other joints. Muscle spasm and attempts to avoid pain distort posture and gait. Prolonged immobility and confinement to the house may cause or aggravate osteoporosis, osteomalacia, obesity, oedema, and incontinence.

Drug treatment may arrest the progress of the disease; physical methods of treatment relieve pain and spasm temporarily. In advanced disease, joint replacement offers the major hope of significant increase of life-space. Surgical replacement of hip and knee joints relieves pain, but may be contraindicated because of multiple joint involvement, cardiorespiratory or renal disease, extreme obesity, brain failure, or depression. Prolonged immobility is also a contraindication, but age alone is not. Pain is not always relieved, and increased activity may precipitate heart failure.

Fracture of the femur

If the fracture occurred while the patient was out of doors, rehabilitation is in most cases rapid and straightforward. If it occurred while the patient was at home, especially in previously housebound patients, rehabilitation is likely to be slow, and its course may be complicated by pressure sores, dehydration, electrolyte disturbance, deep vein thrombosis, infections, delayed wound healing, delayed bony union, and avascular necrosis. A particularly poor functional prognosis is associated with pre-existing incontinence and brain failure.

Parkinsonism

The success of drug treatment, at least before the late stage of the disease is reached, has reduced the difficulty of rehabilitation. However, even with good drug control, symptoms fluctuate throughout the day, and patients feel alternately encouraged and discouraged.

The festinant gait described by Parkinson is not often seen in the aged patient; more often the gait is slow and hesitant. Patients walk better when they walk fast and when they are relaxed. This can be achieved by holding the patient's outstretched hands and walking backwards in front of him or her at a gradually increasing speed.

Short-term improvement has been claimed with the use of breathing exercises, relaxation, distraction, music, talking, rhythmic clapping, startling, or having broad, black and white, transverse stripes on the floor. This last method depends on the generation of a stepping pattern by a rhythmic visual stimulus. It works in laboratory conditions but is not very practicable.

Four-legged walking-frames in most cases increase the difficulty of walking, since akinesis affects the upper as well as the lower limbs, and patients have difficulty in lifting and lowering the frame in the proper sequence. It is much better to use an aid with one or two wheels in the front and two legs at the back. The wheels allow uninterrupted movement, while the legs prevent the aid 'running away' with the patient.

Apraxia of gait

In this condition a normal stepping pattern can be induced temporarily by bringing the weight forward, or by visual stimulation; or it may occur spontaneously. It is not uncommon for the patient to get up and walk to the toilet apparently normally within minutes of being genuinely unable to walk in the presence of the doctor, suggesting that the condition may be an inhibition or suppression of a normal walking pattern.

The gait becomes normal when the centre of mass of the body is brought forward. But patients are unhappy about leaning forward, which seems to make them think that they are about to pitch to the ground, so they quickly revert to leaning back and to the abnormal movements which accompany this position.

Sometimes the condition can be temporarily improved if the therapist walks backwards a fixed distance in front of the patient holding a large bunch of flowers. The patient fixes the gaze on the flowers, which provide a visual frame of reference. This corrects the presumed conflict between the gravitational and the visual vertical.

The forward stance can be achieved, and the fear of falling relieved, by a walking-frame. The disadvantage is that the patient becomes rooted to the frame, and fails to achieve physiological recovery. Frames should be reserved for patients who do not recover after rehabilitation by physiological methods.

Oedema

Severe, persistent oedema of the lower limbs not associated with cardiac failure impairs rehabilitation because of the weight of the leg, the inability to wear shoes, and the tendency to leakage and blistering. Admission to hospital may be necessary for postural treatment, pressure bandaging, the use of alternating pressure, and possibly diuretic drugs.

Amputation

Amputees may have suffered pain, immobility, hope, disappointment, demoralization and a series of operations before the final decision to amputate was taken. A successful outcome may be expected in patients who were in good health before the amputation, in whom the knee joint is preserved, and who are free of postoperative complications. Rehabilitation is more difficult in above-knee amputees and double amputees; and in the presence of diabetes, peripheral neuropathy, ischaemic heart disease, cerebrovascular disease, phantom limb pain, wound infection, burst suture line, a poorly shaped stump, a contracture at the hip joint. These contribute to poor nutrition, impaired balance, reduced exercise tolerance, lack of dexterity, and mental inflexibility. Some patients continue to smoke heavily and ignore warnings that this may endanger the other limb.

Leg amputations are of four main types: above knee, below knee, through knee and Gritti–Stokes. Above-knee amputations heal better than below-knee ones, but cause much more difficulty in rehabilitation, and much greater increase in cardiac work.

If the stump is too long, too short, or badly shaped the artificial limb causes pain and pressure. If the suture line breaks down, healing is slowed and the postamputation prosthetic (PAM) limb, described below, cannot be used. If there is flexion at the hip joint, an artificial limb may cause pressure on the thigh.

Immediately after operation the patient is encouraged to move about in a wheelchair and to learn to transfer, but care is taken that this does not cause flexion of the hip joint or pressure sores.

Walking

The patient may be permitted to hop on the normal limb, supported by a frame or parallel bars, but this overloads the heart and establishes an abnormal walking synergy, which has to be unlearned when an artificial limb is used. It is preferable to commence walking with a temporary limb with an inflatable socket (PAM limb). This can be used for 20 minutes at a time. It promotes balance, confidence, and optimism but some patients find the PAM limb painful and cannot tolerate it. Careful inflation is necessary to avoid ischaemia.

Artificial limbs

Despite improvements in artificial-limb technology, the processes of measuring, forming, modifying, and fitting a limb take time, and may involve many visits to an artificial-limb centre.

The artificial limb is moved by the action of the remaining hip flexors, extensors, abductors and adductors, mainly the iliopsoas and gluteal muscles. The movement of the limb is perceived and monitored by these muscles, and by proprioceptors over the ischium whose previous function was merely to appreciate sitting. The work of walking imposes an excessive load on the heart. It is no easy matter for the elderly patient to learn to interpret the unfamiliar proprioceptive signals, and to activate in correct sequence the different muscles necessary for rising, standing, and walking.

Peripheral neuropathy affecting the upper limbs may impair the ability to pull on and to fasten the limb, and to lock and unlock the knee mechanism. A fall can be catastrophic, and many of the above conditions predispose to falling. Stair climbing is particularly difficult and hazardous.

The details of training the amputee to walk are matters for the specialist physiotherapist.

Notes and references

Life-space

The concept of life-space is described by May *et al.* (1985). Mobility is a major component of self-rated health (Jylha *et al.* 1988).

Physiology

Peripheral age changes, which include decline in muscle mass, reduction in the number of functioning motor units, enlargement of the surviving units, and reduction in the number of fast-twitch fibres (Campbell *et al.* 1973; Kallman *et al.* 1990), affect muscle strength, speed of movement (Larsson *et al.* 1979), and ability to perform complex movements such as rising from a chair (Young 1985).

Movements are also affected by impairment of the central planning and control mechanisms. These are located in the cerebellum, which is the 'function generator' for brisk movements; in the basal ganglia, which generates more gradual movements; in the association areas of the parietal, temporal, and occipital lobes, which take part in the initiation and organization of voluntary movement; and in the motor cortex itself, which represents the final pathway (Desmedt and Godaux 1978). The cerebellum adjusts central motor commands when the voluntary movement is diverted from its intended path by unexpected mechanical conditions. Slowing of neuronal transmission, and lesions at any point in the complex pathway, lead to disturbances of movement, and to delay or failure in correcting unexpected displacement. Programmes of movement are initiated and modified by functional stretch reflexes or 'long loops' (Grimm and Nashner 1978), while stationary neurological deficits represent 'what is left over after compensation is complete'. Gait disturbances in elderly people are attributed in part to 'dyscontrol' in the long-loop system.

Rating scales

Simple rating scales for mobility include the classification of Hoffer *et al.* (1973) into: non-ambulant, limited indoors, unlimited indoors, limited outdoors and unlimited outdoors; and the 'life-space diary' of May *et al.* (1985). Wolfson *et al.* (1990) used a 'gait assessment rating scale' with many components, but found that speed and stride length alone correlated excellently with the total score and with the tendency to fall. Tinetti (1986) and Tinetti and Ginter (1988) found that measures of what patients actually performed were more accurate than what they said they could perform.

Gait analysis

The earlier studies were conducted with photographic methods (Murray *et al.* 1969). Later experiments used mechanical, goniometric, electromyographic, electronic, ultrasonic, and computerized methods (Herman *et al.* 1976; Imms and Edholm 1979, 1981; Nayak *et al.* 1982; Ranu 1987). The value of gait analysis to the clinician has been reviewed by Larish *et al.* (1988).

Fractured femur

The main factors influencing success in rehabilitation are good general medical condition, good social support (Ceder *et al.* 1980), mental clarity, and going out before the fracture (Cobey *et al.* 1976).

Parkinsonism

The slow, shuffling gait of parkinsonian patients is associated with loss of the four-phased pattern of extension and flexion at the knee joint (Frossberg *et al.* 1984). Parkinsonian patients are unable to generate strong bursts of electromyographic activity (Hallet and Khoshbin 1980). Visual stimulation, e.g. by use of horizontal stripes on the floor, increases speed but does not correct the underlying abnormality (Martin 1967). Reduction of anticipatory postural responses slows movement and increases the tendency to fall (Traub *et al.* 1980).

Apraxia of gait

Miller (1986) considers that this term is a misnomer for a condition which lacks some of the essential features of dyspraxia, is not confined to a disturbance of gait, and has additional features of hypokinesia and abnormal reflex activity. In non-demented patients, Fisher (1982) found an association between gait disturbance and the size of the lateral ventricles. Surgical treatment of hydrocephalus improved ambulation in three-quarters of cases (Peterson *et al.* 1985).

Amputation

The number of amputations performed in British hospitals has increased (Sethia *et al.* 1986); but there is a persistently high postoperative mortality (Bodily and Burgess 1983), and high rates of re-amputation and contralateral amputation. This is attributed to 'injudicious attempts at arterial reconstruction'. Better operative results were achieved in special centres (Harrison *et al.* 1987). The postamputation, pneumatic walking-aid (PAM limb) has facilitated early ambulation (Little 1971; Redhead *et al.* 1978; Dickstein *et al.* 1988). Most bilateral amputees learned to walk again, provided that one knee was preserved, that they walked before the second amputation (Steinberg *et al.* 1985), and that their cardiac state allowed them to accept the high energy cost (Waters *et al.* 1976).

References

Bodily, K. C. and Burgess, E. M. (1983). Contralateral limb and patient survival after leg amputation. *American Journal of Surgery*, **146**, 280–2.

Campbell, M. J., McComas, A. J., and Petito, F. (1973). Physiological changes in ageing muscles. *Journal of Neurology, Neurosurgery and Psychiatry*, **36**, 174–82.

Ceder, L., Svensson, K., and Thorngren, K. G. (1980). Statistical prediction of rehabilitation in elderly patients with hip fracture. *Clinical Orthopaedics*, **152**, 185–90.

Cobey, J. C., Cobey, J. H., Conant, L., Weil, V. H., Greenwald, W. F., and Southwick, W. O. (1976). Indicators of recovery from fracture of the hip. *Clinical Orthopaedics*, **117**, 258–62.

Desmedt, J. E. and Godaux, E. (1978). Ballistic skilled movements: load compensation and pathways of the motor commands. *Progress in Clinical Neurophysiology*, **4**, 21–55.

Dickstein, R., Pillar, T., and Mannheim, M. (1988). The pneumatic post ambulation mobility aid in geriatric rehabilitation. *Scandinavian Journal of Rehabilitation Medicine*, **14**, 149–50.

Fisher, M. (1982). Hydrocephalus as a cause of disturbance of gait in the elderly. *Neurology*, **32**, 1358–63.

Frossberg, H., Johnells, B., and Steg, G. (1984). Is Parkinsonian gait caused by a regression to an immature walking pattern? *Advances in Neurology*, **40**, 375–9.

Grimm, R. J. and Nashner, L. M. (1978). Long loop dyscontrol. *Progress in Clinical Neurophysiology*, **4**, 70–84.

Hallet, M. and Khoshbin, S. (1980). A physiological mechanism of bradykinesia. *Brain*, **103**, 301–4.

Harrison, J. D., Southwood, S., and Callum, K. G. (1987). Experience with the skew flap below knee amputation. *British Journal of Surgery*, **74**, 930–1.

Herman, R., Wirta, R., Bamptom, S., and Finley, F. R. (1976). Human solutions for locomotion. I. Single limb analysis. *Advances in Behaviour Biology*, **18**, 13–49.

Hoffer, M. M., Feiwell, E., Perry, J., and Bonnet, C. (1973). Functional ambulation in patients with meningomyelocoele. *Journal of Bone and Joint Surgery*, **55A**, 137–43.

Imms, F. J. and Edholm, O. G. (1979). The assessment of gait and mobility in the elderly. *Age and Ageing*, **8**, 261–7.

Imms, F. J. and Edholm, O. G. (1981). Studies of gait and mobility in the elderly. *Age and Ageing*, **10**, 147–56.

Jylha, M., Leskinen, E., Alanen, E., Leskinen, A-L., and Hoikinnen, E. (1988). Self-rated health and associated factors among men of different ages. *Journal of Gerontology*, **41**, 710–17.

Kallman, D. A., Plato, C. C., and Tobin, J. D. (1990). The role of muscle loss in the age-related decline of grip strength: cross-sectional and longitudinal perceptions. *Journal of Gerontology*, **45**, M82–8.

Larsson, L., Grimby, G., and Karlsson, J. (1979). Muscle strength and speed of movement in relation to age and muscle morphology. *Journal of Applied Physiology*, **46**, 451–6.

Little, J. M. (1971). A pneumatic weight bearing temporary prosthesis for below knee amputees. *Lancet*, **ii**, 271–8.

Martin, J. P. (1967). *The basal ganglia and posture*. Pitman, London.

May, D., Nayak, U. S. L., and Isaacs, B. (1985). The life space diary: a measure of mobility in old people. *International Rehabilitation Medicine*, 7, 182–6.

Miller, N. (1986). Dyspraxia and its management. Croom Helm, Beckenham, Kent.

Murray, M. P., Kory, R. C., and Clarkson, B. H. (1969). Walking patterns in healthy old men. *Journal of Gerontology*, **24**, 169–78.

Nayak, U. S. L., Gabell, A., Simons, M. A., and Isaacs, B. (1982). Measurement of gait and balance in the elderly. *Journal of the American Geriatrics Society*, **30**, 516–20.

Peterson, R. C., Mokri, B., and Laws, E. R. (1985). Surgical treatment of idiopathic hydrocephalus in elderly patients.

Ranu, H. S. (1987). Normal and pathological human gait analysis using miniature triaxial shoe-borne load cells. *American Journal of Physical Medicine*, **66**, 1–11.

Redhead, R. G., Davis, B. C., Robinson, K. D., and Vital, M. (1978). Post amputational pneumatic walking aid. *British Journal of Surgery*, **65**, 611–12.

Steinberg, F. U., Sunwoo, I., and Roettger, R. F. (1985). Prosthetic rehabilitation of geriatric amputee patients: a follow-up study. *Archives of Physical Medicine and Rehabilitation*, **66**, 742–5.

Tinnetti, E. M. (1986). Performance oriented assessment of mobility problems in the elderly patients. *Journal of the American Geriatrics Society*, **34**, 119–26.

Tinnetti, E. M. and Ginter, S. F. (1988). Identifying mobility dysfunction in elderly patients: standard neuromuscular examination or direct assessment? *Journal of the American Medical Association*, **259**, 1190–3.

Traub, M. M., Rothwell, J. C., and Marsden, C. D. (1980). Anticipatory postural reflexes in Parkinson's disease and other akinetic–rigid syndromes and in cerebellar ataxia. *Brain*, **103**, 393–412.

Waters, R. L., Perry, J., Antonelli, D., and Hislop, H. (1976). Energy cost of walking of amputees: the influence of level of amputation. *Journal of Bone and Joint Surgery*, **58A**, 42–51.

Wolfson, L., Whipple, R., Amerman, P., and Tobin, J. N. (1990). Gait assessment in the elderly: a gait abnormality rating scale and its relation to falls. *Journal of Gerontology*, **45**, M12–19.

Young, A. (1985). Exercise physiology in geriatric practice. *Acta Medica Scandinavica* (Suppl.), **711**, 227–32.

Aphorisms

Of mobility

- Movement is to go; mobility is to go places. The best places to go to are on the other side of the street. Mobility is the ability to cross to the other side of the street.
- It takes a child one year to acquire independent movement, and 10 years to acquire independent mobility. An old person can lose both in a day.
- Mobility is vulnerable to diseases of the heart, lungs, muscles, bones, joints, and eyes; and to darkness, weather, and fear.
- Among the causes of staying at home darkness is a sound reason, weather a good excuse, fear a great sadness.

Of life-space

- Life-space diameter at 70 is 10 000 miles, at 80, 200 yards and at 90, 10 feet.
- The words 'take it easy' reduce life-space. The words 'take it steady' increase life-space.
- A little less movement, a lot less life-space; a little more movement, a lot more life-space.

- The border of life-space is, for the dyspnoeic the bottom of a hill, for the dysuric a public toilet, and for the visually handicapped the fall of night.

Of rising and sitting down

- Take care of the chair and rising takes care of itself. Take care of the rise and walking takes care of itself.
- Who sits stands, who sprawls falls.
- Right height chair, straight leg rise. Wrong height chair, angled leg rise.
- No floor is slippery if the legs are straight. All floors are slippery if the legs are not straight.
- Rehabilitation is complete when the patient can sit on a box of eggs without breaking them.
- Walking aids, like drugs, are effective in the short term, toxic in the medium term, and addictive in the long term.

7 *Instability 1: causes, mechanisms, and history*

Introduction

There are few more dismal sights in geriatric medicine than that of an old person crouched over a quadruped walking-frame, lips pursed, eyes on the ground, lurching slowly forward in a melancholy rhythm of lift, lay, shuffle, halt. This picture of sad old age is becoming as uncommon as that of a hemiplegic abducting his leg as he walks, or an incontinent patient lying on a rubber sheet; but when it does occur it represents, as do these other phenomena, a failure to apply physiological principles to the management of the Giants of Geriatrics; and it results in the conversion of an unstable biped into a stable but virtually immobile 'hexapod'.

Terminology and definitions

The balance system of biped man permits him/her to travel at speed through an irregular environment. The following definitions are proposed:

- *Balance* is the set of strategies employed in the maintenance of stability.
- *Normal balance* is the ability to correct an unexpected large displacement in a short time.
- *Instability* is impairment of the ability to correct displacement of the body during its movement through space.
- *Loss of balance* is an ambiguous term. In a single fall, balance is overwhelmed rather than lost. In recurrent falls, balance is impaired.

The words 'dizzy', 'giddy' and 'light headed' describe a sense of instability, which may result from excessive stimulation or incongruous responses of the eyes, the vestibules, or the proprioceptors.

Dizziness may be induced by:

- external linear or rotatory displacement, as in a lift or fairground;
- movement of the head or body, as on rising from a chair or getting out of bed;
- an unfamiliar visual stimulus, e.g. on standing at the edge of a cliff or on top of a ladder.

Dizziness which is experienced at rest, in the absence of these visual and proprioceptive stimuli, is attributable to drugs or disease.

The use of the word 'black-out' may suggest loss of consciousness, or a visual disturbance associated with transient ischaemia of the occipital pole, as occurs in vertebrobasilar insufficiency.

A 'slip' is acceleration and forward extension of the leg, with backward movement of the trunk and head, which is induced by faulty foot placement, low friction of the sole of the shoes, and low friction of the ground surface.

A 'trip' is an unexpected encounter between the moving foot or leg and an unperceived object, and results in retardation and flexion of the advancing limb, with forward movement of the trunk and head.

The term 'drop attack' describes an unexpected and unexplained fall during walking, followed by difficulty in rising. It is doubtful if this is a specific clinical entity, and the term is not used in this book.

Physiology and pathology of balance

Instability is caused by faults in the mechanism for detecting and correcting displacement of the body in space. Information about the position of the body in space flows into the brain from the eyes, the vestibules, and the proprioceptors. This is coordinated, and instructions are despatched to the muscles, which restore or maintain a stable body position.

During movement the eyes subconsciously monitor the position of the body in relation to the environment. The vestibules respond to linear and rotatory accelerations of the head, which are registered by the otolith mechanisms and semicircular canals, respectively. Muscle and joint sense from proprioceptors in the neck, trunk, and limbs are the major source of information about the position and movement of the body.

The information conveyed to the brain from the three sources should be congruent. A fault in one or more sources of information creates uncertainty, delay, bewilderment, and an error in execution. An everyday example of incongruity between visual and proprioceptive information is the erroneous belief of passengers seated in a train at rest in a station that they are moving when an adjacent, stationary train begins to move. In this case, visual information is preferred to proprioceptive.

Sensory misinformation

Sensory misinformation is common in old age because of pathological change in any or all of the sources of sensation. Examples are as follows:

Vision

- Refractive error

- Hyperosmosis of aqueous
- Distortion of lens fibres
- Retinal atrophy
- Cellular loss in occipital lobes

Vestibules
- Fracture and displacement of otoliths
- Leakage of fluid from semicircular canals

Proprioceptors
- Arthritis of hips, knees, ankles
- Joint replacement

Incongruity between sensory messages may induce a sensation of instability and fear of an impending fall.

Balance is not a single function which can be lost. It is a complex of functions which can be impaired in many ways.

Mechanisms of falls

Falls occur when the force and speed of displacement exceed the response of the balance mechanisms. A single fall after a massive displacement may merely imply overload of a healthy balance system; whereas recurrent falls after small displacements imply that the balance system is abnormal. There is no one cause of a fall. Consideration has to be given to the displacement and to the mechanisms for its detection and correction.

The state of the balance mechanism at the time of a fall can be inferred from consideration of the displacement which induced the fall. The smaller the displacement which is associated with the fall, the less likely is it that the fall is physiological.

The displacement of the body may occur:

- without external force during an undistorted planned action, such as rising from a chair or walking on a regular surface;
- as the result of an external force, like a gust of wind on emerging from a high-rise block, or an over-friendly dog jumping up;
- as a consequence of an unexpected encounter during a planned movement with an external hazard, like stepping on a banana skin during walking.

A fall occurring during an undistorted, planned movement is likely to be associated with severe balance impairment. One which results from an unexpected hazard may be due to failure to perceive the hazard, combined with some impairment of balance. A fall from a wobbly ladder may merely indicate imprudence in a subject whose balance is almost normal, as he or she would not otherwise be able to climb the ladder.

Causes of falls

Falls are caused by disturbances in the perception and correction of displacement. The following factors are commonly found in elderly patients who suffer recurrent falls.

Disturbances in the perception of displacement

Visual disorders

- Cataract
- Glaucoma
- Refractive error
- Cortical disorders causing errors of perception of depth and distance

Vestibular disorders

- Vestibular neuronitis
- Benign positional vertigo
- Menière's syndrome
- Otolith degeneration
- Diseases of the semicircular canals
- Brain-stem infarction and ischaemia

Proprioceptive disorders

- Peripheral neuropathy
- Subacute combined degeneration
- Arthritis of cervical, dorsal or lumbar spine
- Spinal cord lesions

Disturbances in the correction of displacement

Central

- Mental slowing
- Cerebral infarction
- Multi-infarct dementia
- Alzheimer's disease, late stages
- Depression

Peripheral

- Acute infections
- Muscle weakness, however caused

- Stroke
- Parkinsonism
- Apraxia of gait
- Multiple sclerosis
- Arthritis
- Maladies of the feet and toe-nails

Systemic
- Malignant disease
- Anaemia
- Malnutrition
- 'Deconditioning'
- Medication, e.g. hypnotics, antidepressants

Falls are sometimes attributed to the following conditions, but it is doubtful what part these diseases really play in the falls:

- vertebrobasilar ischaemia;
- cervical spondylosis;
- carotid sinus syndrome;
- aortic stenosis;
- cardiac dysrhythmia;
- postural hypotension;
- syncope;
- epilepsy.

Vertebrobasilar ischaemia causes transient ischaemic attacks with temporary malfunction of the brain-stem, the cerebellar connections, and perhaps the occipital pole. These occur spontaneously, and rarely on head movement. The symptoms may include vertigo, ataxia, diplopia, and occipital blindness. Falls in the absence of these symptoms should not be attributed to ischaemia.

Cervical spondylosis rarely causes compression of the vertebral artery in the neck on head turning. More commonly it limits turning of the head, and thus interferes with visual surveillance of the environment, e.g. when preparing to cross the street. Vertigo may be induced on head movement because of loss of the proprioceptors normally present in the ligaments and the apophyseal joints of the neck.

Carotid sinus syndrome is rare, and it is not very safe to try to prove its existence by attempting to induce it. It causes loss of consciousness.

Aortic stenosis may cause reduction of cardiac output on effort. The condition may be responsible for recurrent falls on climbing stairs, preceded by transient loss of consciousness.

Cardiac dysrhythmia of sufficient severity to cause asystole induces Stokes–Adams attacks. These should be readily distinguishable from simple falls. The discovery of an arrhythmia in an elderly person who has fallen for no good reason raises the suspicion that the one caused the other.

Postural hypotension

Falls which occur on standing up are often attributed to postural hypotension, especially when a drop in blood pressure is later demonstrated to occur on change of posture.

Many elderly people suffer a drop of blood pressure on change of posture. Many elderly people fall on change of posture. However, in not all of these is the fall due to postural hypotension. A tendency to fall on change of posture should be called *postural instability*. This is common in elderly people with the conditions listed above, and merely means that balance is so severely impaired that the induced displacement of standing up from a seated position cannot be adequately controlled.

A fall in blood pressure of the order of 20 mmHg rarely causes reduction in cerebral blood flow. With very large pressure change on standing, however, blood flow may drop, with resulting dizziness and loss of postural control. These genuine cases of postural instability due to reduction in cerebral blood flow as a consequence of postural hypotension occur most often in elderly patients who are receiving drugs with a hypotensive effect.

Syncope

Many falls are preceded by loss of consciousness or 'black-outs' as patients may call them. The types of syncope likely to be associated with falls in old age include:

- cardiac disease;
- Valsalva manoeuvre, especially after micturition and defaecation;
- vasovagal attacks.

Epilepsy

This is easily identified as the cause of a fall in a known epileptic patient. However, many patients not known previously to suffer from epilepsy experience transient disturbances of consciousness and movement whose description by patients and observers is so incomplete and inconsistent that these episodes are listed in some text books as 'funny turns'. The use of this term in serious medical texts seems an abnegation of the art of clinical description. There is nothing funny about finding oneself lying unaccountably on the floor, perhaps with a broken femur to prove it.

These episodes usually occur in patients with cerebrovascular disease and with non-specific, electro-encephalographic manifestations. Their attribution to epilepsy is difficult to confirm; but when absent states or ill-sustained, convulsive movements are described, a trial of anti-epileptic therapy is justified.

Environmental factors in falls

Environmental hazards cause healthy people to trip or slip, but, unless the displacement is extreme, this does not result in a fall, since the balance mechanism corrects the displacement in time. However, in patients with impaired balance, an unplanned encounter with an environmental feature makes them fall. The factors which may be involved are as follows.

Foot–ground contact
- Misshapen feet
- Badly fitting shoes
- Slippery soles
- A slippery ground surface
- Irregular ground surface

Static hazards
- Changes of ground level, e.g. steps and slopes
- Surface irregularities, e.g. torn carpets, broken paving stones
- Obstacles such as toys, trailing wires

Moving hazards
- Traffic
- People
- Animals
- Furniture on wheels or castors
- Sudden shafts of wind

Visual hazards
- Sudden loss of illumination, e.g. from a light being switched off or a power cut (an alternative meaning for 'black-out')
- Change of illumination on moving from a light to a dark area, e.g. from bright sunshine to shadow
- Dim lighting of critical areas, e.g. at the bottom of a flight of stairs

Instability in old age is most often a manifestation of multiple diseases. The unstable person finds that the normally acceptable discontinuities of the man-made and natural environment have passed beyond the capacity of his or her balance mechanism. He or she faces the threat of falling and takes avoiding action. This may have the paradoxical effect of worsening balance.

Taking the history: falls

An elderly person who has fallen may recall the events poorly, become confused between different episodes, and describe the unfamiliar situation with a limited vocabulary. The facts to be determined from the history are:

- What falls occurred?
- What happened before the fall?
- What happened during the fall?
- What happened after the fall?

What falls occurred?

While some patients give a clear and consistent account of their fall, others may confuse the recent fall with previous falls; or forget that previous falls occurred; or transpose in time or place the events of previous falls. The questioning should concentrate on the time, place, and circumstances of the recent fall before turning to past events. An accurate, consistent history is not always forthcoming. The questioner should accept the limitations of human memory for short, frightening events, and should seek further information from a family member.

What happened before the fall?

The information to be sought includes:

- the state of health before the fall;
- the level of mobility
- location at the time of the fall;
- activity which preceded the fall;
- nature of the displacement.

The location and activity at the time of the fall help to determine the mechanism and prognosis. A fall while ascending stairs may have a different mechanism from one on descending stairs; and a fall in the light is not the same as one in the dark.

Attention should be paid to patients' accounts of their reaction to the fall. With a large displacement, such as falling off a ladder, patients may say 'I was stupid.' They may mean that they were stupid to attempt the task because their balance was not as good as it used to be; or that they believed themselves capable of the task, but made a 'stupid' mistake while performing it.

A patient may say 'I just bent down to pick up the paper . . . and down I went.' These words express unawareness of the balance deficit, and surprise that such a simple action caused a fall.

Patients may answer the question 'What were you doing at the time when you fell?' by saying 'Nothing'. They mean by this that they were doing nothing that seemed likely to cause a fall; and they may add 'I was just walking into the kitchen' or 'I had just got out of bed'.

There may be a clear description of an encounter with an external hazard, or the patient may use one of the following phrases.

'I must have tripped.'
This means I did not trip. I fell for no reason that I can understand, and since trips are thought to cause falls I suppose that 'I must have tripped'. The use of this phrase and of 'I must have slipped' suggest that the balance mechanism is severely impaired.

'My legs (knees) gave way.'
Patients who give this story usually have 'sat down' rather than pitched forward or backward. The cause may be non-compliance of the knee joint as it takes weight in the stance phase of walking. This may be caused by arthritis of the knees or weakness of the quadriceps.

'I missed the chair.'
Patients may misjudge the position of the chair as they sit, indicating an error in depth perception, or misjudgement resulting from anxiety and fear of an imminent fall.

'My feet stuck (or froze).'
This is suggestive of spontaneous arrest of the legs during walking, as in Parkinson's disease or apraxia of gait; but slowly walking patients may fall in this way as a result of catching the sticky sole of the shoe or slipper on a thick-pile carpet.

'I just fell.'
When this is all that the patient can be induced to say the probability is that balance is severely impaired, and that the patient may have become inured to frequent falls.

What happened during the fall?

The statement 'I felt myself going but there was nothing I could do about it' suggests awareness of slowing of corrective reflexes. Patients who did not recognize that they were falling may have been in a state of shock or may have lost consciousness.

The statement 'I saw the ground coming up to meet me' implies that the patient was conscious and fell forward. The relative movement between body and ground is attributed to movement of the ground rather than of the body, reminiscent of the moving-train phenomenon mentioned above.

Victims may fall on the front, back, side, bottom, or knees; or against a wall. They cannot always tell in which direction they fell. The position of soft tissue injuries indicates the direction of the fall.

Forward falls are from tripping, or when there is arrest or freezing. There may be bruising on the temple, indicating that the head has been turned on approaching the ground. The patient with two black eyes after a fall has failed to make this reflex movement, raising the possibility that he or she may be suffering from parkinsonism.

Backward falls occur after a slip, or when there is slowing and muscular weakness. Sideways falls may imply an ineffective attempt to correct the displacement. Falls on the bottom or on the knees occur in patients who say that their 'legs gave way'.

What happened after the fall?

The questioner should determine whether the faller rose by him or herself or whether helped by others; and how long he or she remained on the ground after falling. Among the consequences of falling are:

- 'shock'
- loss of consciousness;
- soft tissue injury;
- fracture;
- 'long lie'.

The word '*shock*', often used by patients, lacks precise medical meaning, but is associated with inactivity and bewilderment. Patients who are alone sometimes say that they lay on the ground after falling for 10 minutes or longer without attempting to rise; but subsequently were able to get up on their own. Others describe fruitless attempts to rise, then suddenly they found themselves on their feet and did not quite know how they had managed to get up. These accounts suggest the existence of a reversible state of impaired voluntary movement after a fall.

Loss of consciousness preceding the fall is suggested when the patient cannot recall having fallen, or, with lesser probability, when he or she says 'I felt myself going'. The use of the phrase 'When I came to . . .' suggests that consciousness was lost, but does not distinguish whether this occurred before or after the fall. The statement that 'I had a black-out' may or may not refer to loss of consciousness.

Soft tissue injury gives a clue to the direction of the fall. Abrasion of the skin of the knees suggests an attempt to crawl. Severe bruising of the face or head suggests that the patient struck a hard object like the edge of the sink on the way to the ground.

Fracture of the upper limb suggests that the patient had an active reflex response to the fall, putting out a hand to save him/herself. Fracture of the hip or pelvis suggests that the protective reflex was absent or delayed.

Proximal hip fractures, especially in thin people, are due to direct impact of the unprotected hip on the ground; or possibly to the bone breaking at its most vulnerable point. In a small proportion of cases of subcapital fracture of the hip, the head of the femur may be avulsed as a result of a sudden, twisting movement, so that fracture precedes the fall. Some patients announce that they heard a bang or felt something give before they fell.

'Long lie'

The term 'long lie' is used arbitrarily to describe a period of more than one hour lying on the floor after a fall; but in many cases the patient lies for 12–24 hours or longer. Long lies are caused by a combination of severe physical and environmental deprivation, and carry a poor prognosis. Patients are too weak, shocked, or mentally impaired to rise spontaneously, and have no means of attracting help, or are unable or unwilling to use such means as they have at their disposal.

Long lies may occur in patients who live with a non-competent person, such as a demented spouse or a schizophrenic son; but usually the victims live alone. Vulnerable patients may or may not have a good communication network, including the telephone and an alarm mechanism. They may or may not be in regular contact with family and neighbours. The circumstances may include excessive use of drugs or alcohol; falling on top of the arms; unwillingness to disturb the neighbours; reluctance to wear a device which activates the alarm system; inability to reach the telephone; poor relations with neighbours; insistence on locking and bolting the doors and windows.

Dizziness

The patient is asked to describe the sensation which he or she experiences, the circumstances in which the symptoms appear, and their duration. Patients may give a clear description of a feeling of instability, rotation, or light headedness; or they may describe mixed sensations; or be unable to go beyond the use of the word 'dizzy'. A circular movement of the hand while the patient describes the symptom may indicate that they experienced rotation; while a sense of instability is sometimes indicated by a wave-like motion of the hand. Dizziness may be induced by movement of the head or by change in body position, or may occur at rest. The relationship to movement may be consistent or inconsistent. A patient who cannot move without experiencing severe dizziness is likely to be suffering from vestibular disease.

Dizziness on moving the head to the side may be due to vestibular or brain-stem or cerebellar disease, or to a visual disorder. Cervical spondylosis may be responsible for some cases of dizziness on turning the head; but the presence of radiological signs of this condition is not a sufficient basis for a cause-and-effect association. In rare cases the vertebral artery in the neck is compressed on head turning by spondylotic outgrowths, but in these circumstances dizziness is accompanied by other neurological signs. Distortion of the visual image as a result of losing or breaking spectacles, or increased sugar content of the aqueous, or following cataract extraction, creates incongruence between vision and proprioception which is interpreted as dizziness.

Dizziness on standing or on rising from bed, usually described as light headedness or a swimming sensation, may be due to reduced blood flow to the brain, or to the experience of instability because of error in the calculation of the muscle power required to attain the upright position. This is characteristic of the experience of rising from bed after an illness. Failure to take account of the weakening of the muscles which has occurred during the illness results in patients not reaching the vertical as they had expected to do. They stagger and feel in danger of falling.

Dizziness which is present at all times and which persists for several weeks and then gradually diminishes is due to vestibular disease. The diminution is a result of cerebral adaptation and suppression. Dizziness due to brain-stem disease tends to be more episodic yet more long lasting. Brief episodes of dizziness occurring during movement are attributable to incongruence between vestibular, proprioceptive, and vestibular information; and inaccurate or inappropriate correction of perceived displacement.

8 *Instability 2: physical examination and rehabilitation*

Physical examination

The patient is examined in bed, then getting out of bed, sitting in a chair, standing up from the chair, standing with eyes open and closed, walking, turning, and sitting down again.

Examination in bed

Examination may reveal disease in many parts of the body which contribute to the tendency to fall. These include:

- distortion of the anatomy of the feet;
- painful corns and bunions;
- arthritis of the ankle and subtalar joints;
- wasting of leg muscles;
- sensory loss or impairment in the legs;
- flaccidity or spasticity of leg muscles;
- clasp-knife or cog-wheel rigidity of leg muscles;
- arthritis of knees and hips;
- wasting of abdominal muscles;
- signs of dehydration;
- irregular heart rhythm;
- aortic stenosis;
- pain or stiffness on moving the neck;
- akinesis of facial muscles;
- spontaneous or positional nystagmus;
- parkinsonian tremor of hands;
- past pointing and other evidence of cerebellar dysfunction.

Examination in a chair

Good sitting balance is determined by the patient's ability to maintain a sitting posture with the back unsupported and the hands on the knees. If he or she is slumped in the chair and cannot sit forward with back unsupported, balance is bad. Stroke patients with impaired balance sit diagonally with the head towards the paralysed side, exposing a triangular area of the back of the chair.

Standing up from a chair

Rising from a chair is achieved in health by an integrated sequence of actions in which the feet press down on the ground, the head and trunk move forward, the knees and hips extend, and the trunk is straightened. In a chair with arms the impetus is increased by downward pressure of the hands on the arms of the chair, extension of the subject's arms, and release of the support.

Optimal start conditions require that:

- The feet are flat on the ground.
- The hips, knees and ankles are at an angle of 90°.
- The back, neck, and head are vertical.
- The eyes look straight forward.
- The seat of the chair is parallel with the ground.
- The back of the chair is perpendicular to the ground.
- The depth of the chair is not greater than the length of the patient's thigh.
- The full length of the patient's back is in contact with the back of the chair.
- The surface of the seat is firm and unyielding.
- The patient's upper arms are vertical.
- The elbows are at an angle of 90–120°.
- The hands are in the same plane as the knees.

Healthy people can achieve a satisfactory rise from less than optimal start conditions. Indeed most chairs supposedly designed for comfort, including the 'easy' chairs commonly found in geriatric wards and the homes of old people, are anything but easy; they should be called 'difficult' chairs. People who find it hard to rise from a chair should not have their difficulties increased by start conditions which increase the mechanical demand upon them.

Rising is impaired by:

- weakness of gluteal and quadriceps muscles;
- limitation of movement at hip, knee, and ankle joints;
- pain in muscles, joints or feet;
- spasticity of lower limbs;
- inco-ordination of trunk and limb muscles;
- akinesia;
- inability to adapt to the start conditions.

Two patterns of difficulty in rising from a chair may be distinguished. In the more common type, which may be called 'flexion failure', patients hoist themselves to a half-way position, with hips and knees partially flexed and hands gripping the arms of the chair, but are unable to complete the action and remain in this position or fall back into the chair. This pattern is found in parkinsonism, arthritis of hips and knees, and general muscle weakness of any cause.

Less commonly, in what may be called 'extension failure', patients adopt a start position with the trunk and lower limbs extended and the feet stretched out in front of the body. The attempt to rise results in the body being pressed into the back of the chair. This condition is observed in patients with severe balance disorders, especially in association with vascular disease of the brain. It is apraxic in nature, since the patient has the capacity to make the necessary movements but is unable to integrate them into a pattern of rising.

Patients with flexion failure may be assisted to stand by optimizing start conditions and by providing a little support in the direction of intended movement. This is best applied by offering counter-pressure on the occiput or hands, and perhaps some pressure on the back, as is described below. Balance testing can then be continued. Patients with extension failure have gravely impaired balance and further testing is not possible.

Examination while standing

Disorders of balance can be detected and roughly quantified by observing the response of the standing subject to positional stress, visual deprivation, and displacement.

Positional stress is applied by asking the subject to stand with feet apart, then with feet together, and finally with one foot in front of the other. Visual deprivation is applied by observing the subject in each position with eyes open and then closed. Displacement may be induced by standing behind the subject and suddenly applying firm pressure or pushing gently on the sternum. The healthy young subject maintains stability in these circumstances. The response of the patient with impaired balance is to sway, stagger, or fall. The relation between the stimulus and the response allows rough quantification of the severity of the balance disturbance.

Sway in the anteroposterior plane is observed by standing at the patient's side and noting the movement of the patient's hair against the background. Sway in the lateral plane is observed by standing behind the patient.

Staggering, in the form of a single step or a series of steps, repositions the body in order to avoid falling. It may be accompanied by throwing the arms out to the side, or by clutching and grabbing at external supports. Patients who stagger forward when they are placed in an upright position seem to feel that they are falling forward, and attempt to correct this by stepping forward. Patients with parkinsonism commonly stagger backwards ('retropulsion') in an attempt to prevent a backward fall.

Falling will occur if staggering fails to correct the imbalance. In cases of severe balance disturbance the patient tends to fall without any corrective movement. The patient is, of course, not allowed to fall.

Examination while walking

Observation of walking is part of the examination of balance. Disordered gait is often due to impaired balance rather than to an intrinsic stepping disorder. The abnormal stepping which occurs in gait disorders due to impaired balance is 'irregularly irregular'.

Patients with balance disturbance tend to flex at hips, knees, ankles, and spine, adopting the 'question mark' or 'chair-shaped' position. This stabilizes them by lowering the centre of mass, but is mechanically and aesthetically unsound.

The provision of a mobile support, like a wheeled walking aid, a case-sheet trolley or a pram, normalizes the stepping pattern of a patient whose gait disorder is due to a balance disturbance but not that of one with an intrinsic stepping disorder.

Arm movement may be absent or irregularly irregular. In mild cases of balance disorder the arms are used efficiently and economically as stabilizers, and shoot out from the side on cornering. In more severe cases, excessive and inefficient use is made of their stabilizing function. In the most severe cases the arms constantly seek an external source of support, and patients clutch and grab at anything or anyone in sight.

Patients with balance disturbance are anxious lest they might fall. Whether as cause or as effect they misjudge distances, and as they approach a chair placed near the end of their walkway they try to sit on it before they have reached it.

Summary of balance disorders

Table 8.1 is a summary of the abnormalities which may be detected during the test.

Clinical presentation

Patients with balance disorders tend to present at departments of geriatric medicine at a late stage, in one of the following ways.

- 'Found lying on floor' (FLOF)
- 'Falling all over the place' (FAOP)
- 'Fractured neck of femur' (FNOF)
- 'Fall back syndrome' (FBS)

Table 8.1 Chief abnormalities found in physical examination of instability

Activity	Minor error	Major error
Stand up	Holds arms of chair too long	Fails to rise Positions feet wrongly Pushes trunk back into chair Falls back into chair Remains 'stuck' half-way up
Remain standing	Visible sway Puts arm out	Excessive sway Staggers Clutches or grabs Falls back
Walk	Hesitates Slow Does not pick up feet Broad base Out toes Puts arm out Does not swing arms	Staggers Stumbles Grabs Rushes Halts Falls Does not 'pass' feet
Turn	Hesitates Slow Puts arms out	Staggers Stumbles Grabs Falls
Sit down		Reaches out before arrival Misses seat Flops down

Found lying on the floor

The patient is usually very old and lives alone. He or she may have fallen during the night on attempting to rise from bed to go to the toilet, was unable to rise or to attract attention, and lay on the floor until discovered.

Falling all over the place

The story is of recurrent falls in a very old person. The patient falls on rising from a chair, or when walking or turning for no apparent reason. Falls occur repeatedly, perhaps several times a day. Injury is unusual.

Fractured neck of femur

Patients with fracture of the proximal femur may be seen on admission to hospital by a geriatrician, who advises on suitability for surgery and helps to plan postoperative management; or they may be referred by an orthopaedic surgeon soon after surgery; or they may not be seen until difficulties in postoperative management become evident, or because an operation was not performed.

The factors responsible for the fall in patients who sustain a fracture after a fall at home are similar to those in patients who do not suffer fracture. Patients who fracture the femur out of doors are a much fitter group and present less often to the geriatrician.

'Fall back syndrome'

The fall back syndrome is seen frequently in patients with impaired balance. These patients have an inveterate tendency to fall back. If brought up into a straight position they immediately step forward. They evidently believe themselves to be upright when they are leaning back; and to be falling forward when they are upright; which is why they step forward to 'prevent' a fall.

It is assumed, but without direct evidence, that patients with this reversible condition have an abnormal setting of the 'posturostat', the mechanism that defines the upright position. This may be induced by prolonged sitting in a low chair with the head and trunk leaning backwards. The condition is sometimes reversed by rehabilitation.

Clinical manifestations

The clinical manifestation of severe balance disorder include:

- the 3F syndrome .... Fear of Further Falling;
- the 3G syndrome .... Grabbing Great Grandmother;
- the 3P syndrome .... Patient with Precocious Parking.

These terms emphasize certain clinical features of advanced balance disorders.

Fear of falling

The life of the patient in the late stages of balance disorder is dominated by fear. The familiar relationship between self and environment is shattered. Visual, vestibular, and proprioceptive clues are in conflict; and each step into unpredictable space is fraught with danger. Patients may be demoralized by fear. This may appear irrational, as when a man who is standing still and appears stable throws out his hands and announces 'I am falling'. Staff often react to this by reassuring the patient and encouraging him to walk; but the patient is not easily reassured, and the extent of his fear has to be recognized.

Grabbing

Clutching and grabbing at every object that appears to offer support, including the coat of any hapless doctor who passes, is a further irrational response to fear. Some patients progress round their homes by clutching at items of furniture. In hospital they may risk falling by grabbing at unstable items like curtains and trolleys.

Precocious parking

This too is in most cases a manifestation of extreme anxiety. Patients are so desperate to regain the safety of a chair that they reach out for it and endeavour to sit down before they have come sufficiently close. This may result in sitting on the floor. Some cases of precocious parking are related to disordered depth perception, as a result of disturbed, central visual function.

The abnormal patterns of balance behaviour which result from fear are potentially reversible if confidence can be restored by successful rehabilitation.

Rehabilitation of patients with balance disturbances

Active rehabilitation comprises:

- treating medical conditions which cause weakness, pain, fatigue, anxiety, or urgency of micturition;
- stopping drugs which cause drowsiness, slowness, or dizziness;
- building muscles to their full strength, and reactions to their full speed, e.g. by ball games while seated;
- ensuring that the patient is dressed and shod properly;
- ensuring that patients and all those who work with them share knowledge of and sympathy with the aims and methods of treatment;
- using correct visual, tactile and proprioceptive signals;
- removing environmental impediments; these comprise:
  —badly designed chairs,
  —ill-fitting and unsupportive footwear,
  —ill-fitting and unsupported clothing,
  —uneven floor surfaces,

—obstructing furniture,
—objects which can be grabbed,
—inadequate illumination,
—ambiguous visual signals, e.g. from shadows,
—anxious people in the vicinity.

The management of patients with balance disorder demands the planned withdrawal of support. The slogan is 'Hands off the patient'. When the hands must be used their primary purpose is not to support the patient but to assist the patient to support him/herself.

The following methods of assisting patients with balance disturbances have been found effective for use by nurses and carers.

Rising from a chair

The hand–head method

This is suitable for patients who can sit up unsupported in the chair. The helper stands at the side of the patient and holds his or her outstretched, clasped hands in one hand while applying counter-pressure on the occiput with the other. When this can be done easily, the hand clasp is changed into a light touch, then support is withdrawn first from the hand, finally from the occiput.

The forearm method

This is suitable for weaker patients who cannot sit unsupported. The helper leans in front of the patient and places his or her forearms under those of the patient, with the patient's hands in the helper's antecubital fossa and the helper's fingers touching the patient's olecranon. The patient is encouraged to push down on the helper's forearms to gain the counter pressure necessary to rise. The helper becomes a dynamic arm-rest, with the opportunity to guide and control the patient and to bring their centre of gravity forward, but without doing any of the work for them.

If the patient clutches the helper's arms he or she is not ready for this method.

The body-hold method

This is used for patients with very poor balance. The helper gets down on his knees in front of the patient and encourages the patient to embrace him round the waist or neck and to bring his body in contact with the helper's body. The patient's legs are grasped between the helper's knees. The helper and the patient rise together by simultaneously extending their knees, while the helper's hands provide counter-pressure on the patient's hips. The patient should do most of the work.

Standing

Patients who cannot stand unsupported should not be supported from behind, since this promotes a tendency to lean back further. Support is given by providing counter-pressure at the pelvis or occiput, and by allowing the outstretched hands to touch but not to grab at the assistant's hands or furniture or wall. The slogan is 'Don't clutch, touch'.

Walking

A fall is to be avoided at all costs, but precautions should be unobtrusive in order to reduce anxiety. If balance is only mildly impaired the patient is encouraged to walk fast with long, confident strides, on the assumption that, as in a riding a bicycle, speed confers stability. The assistant places him/herself behind the patient, where any tendency to lose balance can be immediately corrected.

If a little support is needed this is best given by the helper placing a hand on the patient's occiput to provide counter-pressure and 'steering'. When more support is required, counter-pressure is applied on the small of the back or by placing the hands over the iliac crests. These methods may induce an extensor response in some patients, who force themselves back on to the supporting hands. As an alternative the helper may walk behind the patient and to one side, lightly holding one of the patient's hands, and using touch to speed walking while correcting minor balance deviations.

If this support is insufficient a change of tactics is required. The helper walks backwards in front of the patient and offers the minimum support that is required. This may be:

- a light touch of the palms of the patient's hands on those of the helper;
- supporting the patient's clenched hands, held out in front, on the helper's hands;
- allowing patients to place their hands on the helper's shoulders;
- allowing the patient to grasp the helper round the waist.

Some unstable patients benefit from the use of 'flower power'. The carer walks backwards in front of the patient bearing a large bunch of flowers on which the patient is instructed to fix his or her gaze. The presence of a clear visual frame of reference appears to stabilize the patient.

If walking cannot be accomplished without a great deal of support, recourse has to be made to a walking aid or wheelchair. This decision should be made not too soon and not too late.

Rising from the floor

Patients who are fit to return home, who live alone, who are at risk of further falls, and who are willing and able to learn are taught how to rise from the floor. The recommended sequence is to turn on the side, bend the knees, rise to a kneeling position, and hoist the body on to a stool or low table.

Relatives are instructed how to take the patient through this sequence. They are advised not to attempt to haul the fallen patient by his or her arms. This training should reduce the anxiety of patient and relatives by increasing their self-confidence. It should not be used if it merely heightens anxiety.

Some people who care for patients with balance disturbances prefer to 'help' them by 'walking' them, and by premature and excessive use of a walking-frame. This is not helpful, since what is done for the patient is not done by the patient.

The English verb 'to walk' is intransitive. It is not possible to 'walk' anyone else. 'Walking' the patient consists of two assistants holding the patient's arms and walking with the patient between them. Patients who are 'walked' do not propel themselves forward, nor plan their progression, nor monitor their environment, nor regain normal proprioception. Similarly the frame purchases stability at the cost of mobility and posture. There is a correct time and place for its use.

Walking aids

The correct function of walking aids is to unload painful joints. They are, however, widely used in restoring independent mobility to patients with balance disturbances. Walking aids should provide balance support without loss of the upright posture and the one–two gait rhythm.

A *walking-stick* or cane is useful for stabilizing patients with mild balance disturbances when they are walking out of doors on irregular and unpredictable surfaces. The use of two sticks provides greater stability, but the hands are not free for such tasks as opening doors, when the sticks often fall.

Wheeled aids liberate the stepping mechanism and permit physiological stepping and high speeds. They occupy a lot of floor space in small rooms, are difficult to use on carpet and outdoors, cannot turn corners (especially the four-wheeled varieties), and may 'run away' with the patient.

Triangular aids with three wheels, or a front wheel only and two back-stops, take up the least space and are least difficult to manipulate.

The ubiquitous, four-legged *walking frame* is the most widely used and the least physiological of walking aids. It:

- replaces the upright with the crouched posture;
- replaces forward vision with downward vision;
- replaces rhythmic with start–stop gait;
- replaces swinging arms with lift and thrust;
- requires significantly more energy consumption;
- restricts movement in confined spaces;
- prevents the use of the hands in carrying, lifting, and manipulating;
- makes the arms rather than the legs the major channel of proprioception.

On other hand the four-legged frame is light and easily portable and gives a great sense of stability. In unstable patients this last advantage may outweigh its disadvantages.

Sending the patient home

Patients who remain at risk of falling despite rehabilitation cause great anxiety to relatives. Instability is a common reason for admission to residential and nursing homes.

A predischarge, home-assessment visit is recommended at which the patient and relatives and members of the rehabilitation team plan how further falls can be avoided and what to do if they occur. These aims are achieved by:

- training relatives and patients in methods of maintaining balance and in rising after a fall;
- making sure that the chair, the bed, and the toilet seat are of the right height and design;
- providing grab handles in the bathroom;
- ensuring that footware and floor surfaces are safe;
- diminishing clutter;
- ensuring that lighting is adequate and is properly used;
- identifying sources of help in the event of a further fall, e.g. a neighbour, and a method of making contact;
- ensuring the patient's agreement to contacting the neighbour or warden;
- arranging an 'informal alarm', i.e. an agreed method of attracting attention, like banging on the wall, in the event of need;
- installing alarm devices or telephones and ensuring that the patient can use them;
- ensuring that someone telephones or visits at least once every 24 hours;
- leaving the key with a neighbour;
- ensuring that the door is not bolted or blocked;
- leaving the telephone number of a member of the rehabilitation team.

Personal alarms

Personal alarms are installed in sheltered housing, retirement housing, and many private houses. These are activated by the patient pulling a cord or touching a switch on a wrist or body-worn device. They alert a warden, a family member, or a central control room, who can then communicate with the patient by voice or telephone, or come to see what is wrong.

Personal alarms would be of special value in patients at risk of further falls were it not for the vagaries of human nature. Among the reasons for their failure to solve the problem are:

- There are no string cords where the patient is lying.
- The patient has tied up the string cords to prevent their mischievous use by children or because they are considered unnecessary and obtrusive.
- The patient is not wearing the pendant or wrist band.
- The patient is lying on top of these and cannot reach them.
- The patient does not want to disturb others.
- The patient is wedged against the door.

Conclusion

Instability is a typical 'Giant', which causes great misery and anxiety to large numbers of elderly people. It is rarely due to a single cause and it cannot often be cured. On the other hand it offers to the doctor many opportunities for identifying and modifying correctable factors, and it responds well to rehabilitation.

The heading 'Falls' did not until recently appear in the indexes of medical textbooks or the rubrics of databases. The subject is now a major field of developing knowledge in the practice of geriatric medicine.

Notes and references

Reviews

Among reviews of the subject are those of Radebaugh *et al.* (1985) and Gibson *et al.* (1987).

Definition

Variation in the prevalence of falls in different studies may be related to differences in definition and methods of recording. Sheldon (1960) asked people whether they had 'a tendency to fall', which may have identified only those who suffered multiple falls. Blake *et al.* (1988) asked subjects how often they had fallen in the past year; and may have selected only those with a good memory.

Sudden, unexplained falls during movement were labelled 'drop attacks' by Sheldon (1960), who speculated that they might be due to loss of cerebellar neurones. Evans (1988) doubted whether they constituted a distinct clinical category, and attributed them to quadriceps weakness and instability of the knee, causing 'buckling'.

Miller (1986) considered that so-called apraxia of gait lacked the essential features of apraxia, was not confined to a disturbance of gait, and had additional manifestations of hypokinesia and abnormal reflex activity.

Neurophysiology

A sudden, potentially destabilizing movement is preceded by activation of postural muscles in anticipation of the adjustment required to maintain equilibrium (Cordo and Nashner 1982). Similarly, a step down from a height is accompanied by 'preprogrammed' contraction of the soleus muscle in order to soften the impact (Greenwood and Hopkins 1976). The slowing of these responses in old people (Inglin and Woolacott 1988) contributes to the liability to fall on unexpected displacement; while the reduction of anticipatory postural responses contributes to the inability of parkinsonian patients to adjust to unexpected displacements (Traub *et al.* 1980).

Clement *et al.* (1984) suggested that balance comprises a 'conservative' process, giving a stable reference system or 'static balance'; and an 'operative' process, which is rapidly responsive to unexpected perturbations. Even in the presence of good static balance, the operative process may not be fast or accurate enough to correct an unexpected displacement or a faulty sensory perception.

Pathology

Pathological changes in the vestibules which contribute to balance disturbances by sending false signals include: degenerative changes in the neuroepithelium of the vestibules (Johnsson 1971; Rosenhall 1973; Schuknecht *et al.* 1965); and degeneration and dropping off of the otoconia (Schuknecht 1969; Ross *et al.* 1976).

Aetiology of falls

Compared with old people who have not fallen, those who have fallen show: decreased physical activity; increased sway; diminished proprioception (Brocklehurst *et al.* 1982); multiple sensory deficits (Drachmann and

Hart 1972; Sharma and MacLennan 1988); poor near vision (Lichtenstein *et al.* 1988); slow gait, short stride length, and other gait abnormalities (Wolfson *et al.* 1990; Lichtenstein *et al.* 1990); lack of recreational exercise (Farmer *et al.* 1989); arthritis of the lower limbs; and greater use of drugs (Campbell *et al.* 1989). The more of these adverse factors that are present, the greater is the likelihood of falling (Tinnettii *et al.* 1986).

Vision makes a major contribution to balance (Dornan *et al.* 1973; Lee and Lishman 1975; Nashner and Berthoz 1978; Manchester *et al.* 1988). This applies to focal or near vision (Lichtenstein *et al.* 1988); and to ambient vision, which is dependent on the peripheral retina and its central connections (Paulus *et al.* 1984; Leibowitz and Shubert 1985). Age impairs the ability to identify visually the direction of a movement (Warren *et al.* 1990).

Cardiovascular factors

Postural changes in pulse rate and in blood pressure, although relatively common in ill, old people and those on medication (Johnson *et al.* 1965; Caird *et al.* 1973) are less in well, old people than in young people; and are rarely accompanied by symptoms (Goldstein and Shapiro 1990). Disturbances of cardiac rhythm are common in elderly people, but the evidence linking them with dizzy attacks and falls is weak (Taylor and Stout 1983).

Tests of balance

Experimental methods of studying balance include the measurement of postural sway on standing (Sheldon 1963; Overstall *et al.* 1977); and on deprivation of visual and proprioceptive information by closing the eyes, standing on foam, or exposure to illusory movement (Lee and Lishman 1975; Ring *et al.* 1988*a*, *b*, 1989). Other methods include study of the electrical activity of muscles in response to the displacement of a supporting surface (Diener *et al.* 1982; Era and Heikkinen 1985); and observing the response to a jerking displacement caused by the drop of a weight on a pulley (Wolfson *et al.* 1986).

In clinical practice, balance is tested by observing the standing patient's response to closing the eyes or tapping the chest (Wild *et al.* 1986); by asking him to stand 'long stride' or on one foot (Tinetti 1986); or to rise, stand, walk, turn, and sit again (Mathias *et al.* 1986). Tinetti *et al.* (1986) derived a balance and gait index from a number of clinical observations and bedside tests, and found that this correlated well with more complex tests of balance (Lichtenstein *et al.* 1990) and with the probability of falling (Tinetti 1986).

Dizziness

Brandt and Daroff (1980) defined vertigo as 'displeasing distortion of static gravitational orientation; or erroneous perception of either self or object motion'. Sloane *et al.* (1989) distinguished five usages of the word 'dizzy': a spinning sensation, a disturbance of equilibrium, light headedness, a combination of these sensations, and 'a feeling that is difficult to articulate'.

Drachmann and Hart (1972) exposed patients who complained of dizziness to different types of motion. Patients were able to match their symptoms to the motion, but there was no relation between symptoms and pathology.

Fractures

After the age of 75 the incidence of forearm fractures ceases to rise with age (Miller and Evans 1985) or declines (Garraway *et al.* 1979; Owen *et al.* 1982), while that of fracture of the hip continues to rise steeply. The person reaches the ground and strikes the hip before he or she can put out a hand for protection. This is attributed to slowing of protective reflexes (Larsson *et al.* 1979), poor orientation, lack of local shock absorbence, and decreased bone strength (Cummings and Nevitt 1989).

Prognosis

While single falls carry little risk (Evans 1988), those severe enough to be brought to the attention of a doctor (Wild *et al.* 1981), and especially those associated with a prolonged period spent on the floor after falling (Murphy and Isaacs 1982), carry a high mortality over the succeeding year, presumably as a result of the diseases associated with the fall. Patients who have suffered frequent falls experience a deterioration of gait pattern, loss of confidence in balance (Murphy and Isaacs 1982), and reduction of physical activity (Vellas *et al.* 1987). However the large number of factors involved in falls offers many opportunities for prevention (Gibson *et al.* 1987).

References

Blake, A. J. *et al.* (1988). Falls by elderly people at home: prevalence and associated factors. *Age and Ageing*, 17, 365–72.

Brandt, T. and Daroff, R. B. (1980). The multisensory physiological and pathological vertigo syndromes. *Annals of Neurology*, 7, 195–203.

Brocklehurst, J. C., Robertson, D., and James-Groom, P. (1982). Clinical correlates of sway in old age: sensory modalities. *Age and Ageing*, **11**, 1–10.

Caird, F. J., Andrews, G. R., Kennedy, R. D. (1973). Effect of posture on blood pressure in the elderly. *British Heart Journal*, **35**, 527–30.

Campbell, A. J., Borrie, M. J., and Spears, G. F. (1989). Risk factors for falls in a community based prospective study of people 70 years and older. *Journal of Gerontology*, **44M**, 112–17.

Clement, G., Gurfinkel, V. S., Lestienne, F., Lipshits, M. I., and Popov, K. E. (1984). Adaptation of postural control to weightlessness. *Experimental Brain Research*, **57**, 61–72.

Cordo, P. J. and Nashner, L. M. (1982). Properties of postural adjustments associated with rapid arm movements. *Journal of Neurophysiology*, **47**, 287–302.

Cummings, S. R. and Nevitt, M. C. (1989). A hypothesis: the causes of hip fracture. *Journal of Gerontology*, **44M**, 107–11.

Diener, H. C., Dichgans, J., Bruzek, W., and Selinka, H. (1982). Stabilization of human posture during induced oscillations of body. *Experimental Brain Research*, **45**, 126–32.

Dornan, M. B., Fernie, G. R., and Holliday, D. J. (1973). Visual input: its importance in the control of postural sway. *Archives of Physical Medicine and Rehabilitation*, **59**, 586–91.

Drachmann, D. A. and Hart, C. W. (1972). An approach to the dizzy patient. *Neurology*, **22**, 323–34.

Era, P. and Heikkinen, E. (1985). Postural sway during standing and unexpected disturbance of balance in men of different ages. *Journal of Gerontology*, **40**, 287–95.

Evans, J. G. (1988). Falls and fractures. *Age and Ageing*, **17**, 361–4.

Farmer, M. E., Harris, T., Madans, J. H., Wallace, R. B., Cornom-Huntley, J., and White, L. R. (1989). Anthropometric indicators and hip fracture in the NHANES. I. Epidemiology follow-up study. *Journal of the American Geriatrics Society*, **37**, 9–16.

Garraway, W. M., Stauffer, R. N., Kurland, L. T., and O'Fallon, W. M. (1979). Limb fracture in a defined population. I: Frequency and distribution. *Mayo Clinic Proceedings*, **54**, 701–7.

Gibson, M. J., Andres, R. O., Isaacs, B., Radebaugh, T., and Worm-Petersen, J. (1987). The prevention of falls in later life: report of the Kellog International Work Group on the prevention of falls by the elderly. *Danish Medical Bulletin*, **34** (Suppl.), 1–24.

Goldstein, I. B. and Shapiro, D. (1990). Cardiovascular responses during postural change in the elderly. *Journal of Gerontology*, **45M**, 20–5.

Greenwood, R. and Hopkins, A. (1976). Landing from an unexpected fall and a voluntary step. *Brain*, **99**, 375–86.

Inglin, B. and Woollacott, M. (1988). Age-related changes in anticipating postural adjustments associated with arm movements. *Journal of Gerontology*, **43M**, 105–13.

Johnson, R. H., Smith, A. C., Spalding, J. M. K., and Wollner, L. (1965). Effect of posture on blood pressure in elderly patients. *Lancet*, **i**, 731–3.

Johnsson, L. G. (1971). Degenerative changes and anomalies of the vestibular system in man. *Laryngoscope*, **81**, 1682–94.

Larsson, L., Grimby, G., and Karlsson, J. (1979). Muscle strength and speed of movement in relation to age and muscle morphology. *Journal of Applied Physiology*, **46**, 451–6.

Lee, D. N. and Lishman, J. R. (1975). Visual proprioceptive control of stance. *Journal of Human Movement Studies*, **1**, 87–95.

Leibowitz, H. W. and Shupert, C. L. (1985). Spatial orientation mechanisms and their implications for falls. *Clinics in Geriatric Medicine,* **1,** 571–80.

Lichtenstein, M. J., Shields, S. L., Shiavi, R. G., and Burger, M. C. (1988). Clinical determinants of biomechanical platform measures of balance in aged women. *Journal of the American Geriatrics Society,* **36,** 996–1002.

Lichtenstein, M. J., Burger, M. C., Shields, S. L., and Shiavi, R. G. (1990). Comparison of biomechanics platform measures of balance and videotaped measures of gait with a clinical mobility scale in elderly women. *Journal of Gerontology,* **45,** M49–54.

Manchester, D., Woollacott, M., Zederbauer-Hilton, N., and Marin, D. (1988). Visual, vestibular and somatosensory contributions to balance control in the older adult. *Journal of Gerontology,* **44,** M118–127.

Mathias, S., Nayak, U. S. L., and Isaacs, B. (1986). Balance in elderly patients: the Get up and Go test. *Archives of Physical Medicine and Rehabilitation,* **67,** 387–9.

Miller, N. (1986). *Dyspraxia and its management.* Croom Helm, Beckenham, Kent.

Miller, S. W. M. and Evans, J. G. (1985). Fractures of the distal forearm in Newcastle: an epidemiological survey. *Age and Ageing,* **14,** 155–8.

Murphy, J. and Isaacs, B. (1982). The post-fall syndrome. *Gerontology,* **28,** 265–70.

Nashner, L. and Berthoz, A. (1978). Visual contribution to rapid motor response during postural control. *Brain Research,* **150,** 403–7.

Nashner, L. M., Black, F. O., and Wall, C. (1982). Adaptation to altered support and visual conditions during stance: patients with vestibular deficits. *Journal of Neuroscience,* **2,** 536–44.

Overstall, P. W., Exton-Smith, A. N., Imms, F. J., and Johnson, A. L. (1977). Falls in elderly related to postural imbalance. *British Medical Journal,* **1,** 261–4.

Owen, R. A., Melton, L. J., Johnson, K. A., Ilstrup, D. M., and Riggs, B. L. (1982). Incidence of Colles fracture in a North American community. *American Journal of Public Health,* **72,** 605–7.

Paulus, W. M., Straube, A., and Brandt, T. (1984). Visual stabilization of posture: physiological stimulus characteristics and clinical aspects. *Brain,* **107,** 1143–63.

Radebaugh, T. S., Hadley, E., and Suzman, R. (1985). Falls in the elderly: biologic and behavioural aspects. *Clinics in Geriatric Medicine,* **1,** 497–697.

Ring, C., Matthews, R., Nayak, U. S. L., Isaacs, B. (1988*a*). Visual push: a sensitive measure of dynamic balance in man. *Archives of Physical Medicine and Rehabilitation,* **69,** 256–60.

Ring, C., Nayak, U. S. L., and Isaacs, B. (1988*b*). Balance function in elderly people who have and who have not fallen. *Archives of Physical Medicine and Rehabilitation,* **69,** 261–4.

Ring, C., Nayak, U. S. L., and Isaacs, B. (1989). The effect of visual deprivation and proprioceptive change on postural sway in healthy adults. *Journal of the American Geriatrics Society,* **37,** 745–9.

Rosenhall, U. (1973). Degenerative patterns in the aging human neuroepithelium. *Acta Otolaryngologica* (Stockholm), **76,** 208–20.

Ross, M., Johnsson, L. G., Peacor, D., and Allard, F. (1976). Observations on normal and degenerating human otoconia. *Annals of Otolaryngology,* **85,** 310–26.

Schuknecht, H. F. (1969). Cupulolithiasis. *Acta Otolaryngologica,* **90,** 765–8.

Schuknecht, H. F., Iganashi, M., and Gaceck, R. R. (1965). The pathological types of cochleosaccular degeneration. *Acta Otolaryngologica,* **59,** 154–70.

Sharma, J. C. and MacLennan, W. T. (1988). Causes of ataxia in patients attending a falls laboratory. *Age and Ageing*, **17**, 94–102.

Sheldon, J. H. (1960). On the natural history of falls in old age. *British Medical Journal*, **2**, 1685–90.

Sheldon, J. H. (1963). The effect of age on the control of sway. *Gerontologia Clinica*, **5**, 129–38.

Sloane, P., Blazer, D., and George, L. K. (1989). Dizziness in a community elderly population. *Journal of the American Geriatrics Society*, **37**, 101–8.

Taylor, I. C. and Stout, R. W. (1982). Is ambulatory electrocardiography a useful investigation in elderly patients with 'funny turns'? *Age and Ageing*, **12**, 211–16.

Tinetti, E. M. (1986). Performance oriented assessment of mobility problems in the elderly patients. *Journal of the American Geriatrics Society*, **34**, 845–50.

Tinetti, M. E., Williams, T. F., and Mayewski, R. (1986). Fall risk index for elderly patients based on number of chronic disabilities. *American Journal of Medicine*, **80**, 429–34.

Traub, M. M., Rothwell, J. C., and Marsden, C. D. (1980). Anticipatory postural reflexes in Parkinson's disease and other akinetic–rigid syndromes and in cerebellar ataxia. *Brain*, **103**, 393–412.

Vellas, B., Cayla, F., and Bocquet, H. (1987). Prospective study of restriction of activity in older people after falls. *Age and Ageing*, **16**, 189–93.

Warren, W. H., Blackwell, A. W., and Morris, M. W. (1990). Age differences in perceiving the direction of self motion for optical flow. *Journal of Gerontology*, **44**, P147–53.

Wild, D., Nayak, U. S. L., and Isaacs, B. (1981). How dangerous are falls in old people? *British Medical Journal*, **282**, 266–8.

Wild, D., Nayak, U. S. L., and Isaacs, B. (1981). Description, presentation and prevention of falls in old people at home. *Rheumatology and Rehabilitation*, **20**, 153–9.

Wolfson, L., Whipple, R., Amerman, P., and Kleinberg, A. (1986). Stressing the postural response. *Journal of the American Geriatrics Society*, **34**, 845–50.

Wolfson, L., Whipple, R., Amerman, P., and Tobin, J. N. (1990). Gait assessment in the elderly: a gait abnormality rating scale and its relation to falls. *Journal of Gerontology*, **45**, M12–19.

Aphorisms

Of balance

- Balance enables man to move dangerously.
- Balance corrects unplanned displacement.
- Balance is like the Treasury; its strength is its hidden reserves.
- Quadrupeds are geometrically stable; bipeds are geometrically unstable.
- Balance converts geometric instability into functional stability.
- Balance has three sensory inputs: visual, proprioceptive, and vestibular; and the greatest of these is proprioception.

Of imbalance

- Congruity between sensory inputs is stability; incongruity between sensory inputs is instability.
- Good balance is like a good prime minister, who resolves conflicts between Departments. Poor balance is like a poor prime minister, who cannot resolve conflicts between Departments.
- Dizziness occurs when the postural senses disagree and the brain does not know which to believe.
- Dizziness results from the inappropriate correction of a misperceived displacement.
- In late life, sensory error increases and balance reserve diminishes.
- Sensory systems which make mistakes issue conflicting instructions. Muscles which receive conflicting instructions make mistakes. That is why old people fall.
- When vision and proprioception disagree we believe vision and disbelieve proprioception. That is dizziness.
- When vision and vestibule disagree we believe vestibule and disbelieve vision. That is vertigo.
- Misplacement is born from misinformation out of misperception, and sires a fall.
- Visuospatial mismatch during movement causes fear of falling.
- Impaired balance in late life is expressed in the 3F, the 3G, the 3P and the FB syndromes. The 3F syndrome is Fear of Further Falling. It inhibits gait, shortens the step and removes arm swing. The 3G syndrome is the Grabbing Great Grandmother. The patient clutches any object in sight, including the doctor. The 3P syndrome is the Patient with Precocious Parking who tries to sit on a chair that has not yet been reached. The FB syndrome is the Fall Back Syndrome. When the patient is moved into an upright position he inexorably steps forward. He has a faulty reading on his posturostat.

Of falls

- Falls result from non-correction of displacement, and from correction of non-displacement.
- Non-correction of displacement is due to failure to perceive its occurrence.
- Correction of a non-displacement is due to failure to perceive its non-occurrence.
- Falls occur when the speed and extent of displacement exceed the speed and accuracy of correction.

- One millisecond of delay at each synapse is the difference between falling and not falling.
- A bruise on the wrist means an active arm reflex; a bruise on the temple means an active neck reflex; two black eyes means inactive reflexes.
- You can tell a parkinsonian by his two lovely black eyes.
- The incidence of fracture of the upper limb increases in youth: that is exploration.
  It diminishes in adulthood: that is maturation.
  It increases again in late life: that is retardation.
  It diminishes again in extreme old age: that is degeneration.
- In exploration, displacement exceeds balance.
  In maturation, balance exceeds displacement.
  In retardation, displacement again exceeds balance.
  In degeneration, displacement is halted by immobility.

Of rehabilitation

- The danger in walking, as in flying, is at take off and landing. The rest looks after itself.
- If a patient can rise from a chair and stand unsupported s/he can learn to walk. If a patient cannot rise from a chair and stand unsupported s/he cannot learn to walk.
- Rehabilitation of balance corrects sensory error, speeds central activity, and strengthens muscles.
- To 'walk' a patient is grammatically and physiologically impossible. To 'drag' a patient is grammatically and descriptively accurate.
- Walking, like swimming and cycling, cannot be done for one person by another.

Of frames

- There are few more dismal sights in geriatric medicine than that of an old man crouched over a quadrupod walking-frame, eyes on the ground, lurching slowly forward in a melancholy rhythm of lift, lay, shuffle, halt.
- The number of legs is directly proportional to stability and inversely proportional to mobility. A quadrupod frame makes a patient three times as stable and one-third as mobile.
- The frame replaces the upright posture with the crouched; the rhythmic motion of the legs with dysrhythmia; and the swing of the arms with lift and thrust.

- The frame makes the arms rather than the legs the major channel of proprioception.
- Frames have a place in rehabilitation of balance disorders, but only after all else has failed, and sometimes not even then.

9 *Incontinence*

Introduction

The Giant Incontinence was once an example of the Inverse Care Law, which states that the more common a condition is the less attention is devoted to it. There is now considerable professional interest and public awareness; but in some places the gap between knowledge and practice remains.

Definitions

Incontinence

Incontinence is the involuntary and inappropriate passage of urine and/or faeces. Incontinent patients pass urine or faeces into their clothing or bedding or chair; or they dribble it on to the floor on their way to the toilet.

Incontinence results from a combination of impaired function of the bladder and bowel; and one or more extrinsic factors. These include:

- disturbances of the urinary tract, including infection, outlet obstruction, anatomical weakness of the sphincter;
- impaired mobility and dexterity;
- constipation;
- hypnotic, sympathomimetic and antiparasympathetic drugs;
- brain failure;
- an adverse physical or psychological environment.

Impaired bladder and bowel function do not of themselves necessarily cause incontinence, but may be responsible for urgency, precipitancy, and dysuria.

The inappropriate passage of urine and faeces may occur in patients with brain failure who have normal bladder and bowel function. They may pass urine voluntarily against the wall or into a flower pot, in the mistaken believe that this is an appropriate thing to do.

Many different terms are used to describe and classify incontinence. An attempt has been made by the International Continence Society to rationalize and standardize these. Their approach will be followed here, with a few exceptions.

Mechanisms of incontinence of urine

Incontinence of urine results from the interplay of intrinsic and extrinsic factors. The *intrinsic factor* is destabilization of the mechanism of bladder emptying, which results from physiological and pathological changes in kidney and bladder function. The *extrinsic factors* include:

- the quality and quantity of the urine;
- the state of the bowel;
- mobility;
- dexterity;
- state of alertness;
- intellectual status;
- emotional status;
- the physical environment;
- the psychological environment.

Normal bladder function

The healthy young person with normal fluid intake secretes urine during the day at an average rate of about 100 ml per hour, and rather less during the night.

Awareness of the state of the bladder is signalled in three stages. The first sensation of fullness occurs at a volume of about 300 ml, i.e. after three hours when fluid intake is average. The sensation is mild and easily suppressed. A sense of discomfort occurs at a volume of about 600 ml, i.e. after six hours when fluid intake is average. The bladder feels full, and the sensation fluctuates from mild to severe discomfort. At a volume of 750 ml or more the subject experiences an irresistible desire to micturate, feels 'desperate', and may leak urine.

The sensations of discomfort are associated with contractions of the detrusor muscle of the bladder. Escape of urine is prevented by the closure mechanism, which is more effective in males than in females. The bladder contractions are inhibited by descending impulses from the frontal cortex. The pressure within the bladder does not rise above 15 cm of water.

Intrinsic changes

In later life the circadian rhythm of urine secretion changes, with more urine being secreted at night than by day. The following changes have been observed in bladder function in both continent and incontinent elderly people. Some are clearly pathological, and are associated with frequency, urgency, and perhaps incontinence. Others may represent physiological age changes, and do not necessarily provoke symptoms.

- Bladder sensation is dulled.
- The detrusor muscle contracts when the volume of urine in the bladder is comparatively low.
- The intensity of the detrusor contractions increases, and bladder pressure may rise above 15 cm of water.
- Voluntary inhibition of the detrusor contractions is delayed and diminished.
- Uninhibited bladder contractions may occur on change of posture, e.g. on rising from a chair.
- The mechanism of closure of the bladder outlet may be inefficient, especially in parous women with damage to the birth canal.
- The outflow may be obstructed, e.g. by an enlarged prostate or a loaded bowel.
- Bladder emptying may be incomplete as a result of poor detrusor compliance, the action of sympathomimetic drugs, and outflow obstruction.

Mechanisms of incontinence

These changes can lead to incontinence by four main mechanisms:

1. High-pressure, involuntary detrusor contractions overcome the closure mechanism of the sphincter.
2. An incompetent sphincter mechanism allows the passage of urine at relatively low bladder pressure.
3. Obstruction of the outlet allows the accumulation of large quantities of urine in the bladder, until overflow leakage or emptying occurs.
4. An inert detrusor muscle fails to contract despite the accumulation of a large volume of urine in the bladder, until overflow leakage or emptying occurs.

In many elderly, incontinent patients more than one mechanism operates. It is sometimes possible to identify which is responsible for urinary tract dysfunction from the clinical findings alone, or by measuring the rate of urine flow from the full bladder and the pressure in the bladder at different volumes. These urodynamic studies necessitate inserting catheters into the bladder and bowel, and instilling saline into the bladder. The main value of urodynamic studies is in patients in whom surgical treatment is contemplated. In elderly patients with established incontinence the correlation between clinical, and urodynamic findings is not close.

Extrinsic factors

The following extrinsic factors increase the likelihood of a patient with intrinsic bladder changes becoming incontinent.

1. Excessive urine flow, as in diabetes, renal failure, the use of diuretic drugs and excessive fluid consumption.
2. Urinary tract infection.
3. Urinary calculus or tumour.
4. A loaded rectum.
5. Reduced mobility, which increases the time between experiencing the desire to micturate and being able to do so.
6. Osteoarthritis of the hips, preventing the adoption of a squatting position on the toilet and causing incomplete bladder emptying.
7. Loss of dexterity, resulting from stroke, arthritis, Parkinson's disease, Dupuytren's contraction, physiological tremor, peripheral neuropathy, etc., and increasing the time taken to loosen clothing.
8. Diminished alertness as a result of altered sleep pattern, lack of physical and psychosocial stimulation, and the use of hypnotic and sedative drugs.
9. Intellectual impairment, which diminishes awareness of the need to micturate, of the time since last micturition and of the location of toilet facilities.
10. Dementia and depression, which diminish self-respect and weaken the desire to remain continent. Shame may lead to secretiveness and denial; or else shamelessness and indifference may be prominent.
11. Anxiety, which increases frequency and urgency.
12. Behaviour disorders, sometimes unkindly described as 'senile delinquency', in which the patient is suspected of using the symptom of incontinence as a stick with which to beat carers, or as a banner on which to blazon the invalid state. The existence of this mechanism is speculative; and if it does exist, it is probably uncommon.
13. The physical environment, notably the position of the toilet, its distance from the patient's normal place, its appearance, temperature, odour, privacy; the height of the seat; the support offered.

The psychology of incontinence

Incontinence represents regression to childhood behaviour, and creates in those who suffer from it a sense of shame, guilt, and failure. This may not be dispersed even by the kindest and most sympathetic of carers. Cleaning up after an incontinent patient is unpleasant. Most carers perform the task discreetly, and try not to convey to the patient any sense of blame. They give active encouragement, and reward the patient for success in remaining dry. This behaviour strongly favours reduction in episodes of incontinence.

On the other hand any tendency to treat incontinent patients as a source of trouble, or as guilty of a preventable misdemeanour, even by the use of such innocent questions as 'Why didn't you call me earlier?' may aggravate depression; or may even create unconcern or indifference. Patients can easily be reduced to tears or provoked to anger by a hasty word. Creation and maintenance of the right psychological environment for the management of a ward with many incontinent patients requires angelic patience and generous staffing.

Taking the history

The questions which are asked in taking a bladder and bowel history should not embarrass the patient, should not be difficult to answer, and should provide information which assists assessment of the nature, duration, severity, and cause of the incontinence, and of the patient's reaction to the symptoms.

The conventional opening question 'Any trouble with the waterworks?' may produce bewilderment or denial. An alternative opening approach is to enquire not about the incontinence itself but about its consequences, e.g.:

- increased need for laundering sheets and clothing;
- increased need for baths;
- uriniferous odour from clothing and carpets;
- reluctance to go out for trips to shops, church, pub;
- reluctance to visit the homes of others;
- reluctance to invite others to the house;
- puddles on the floor;
- voluntary reduction of fluids, especially in the evening;
- waking to find the bed wet;
- anxiety lest an 'accident' should occur;
- anxiety lest a carer should be annoyed;
- acquiring rags, towels, toilet rolls, rubber sheets, sanitary pads or incontinence pads;
- replying to mail-order advertisements for pads;
- telling (or not telling) the GP about the problem.

This can then be followed by more direct questions about the nature and duration of the incontinence. The question 'How long have you been incontinent?' fails to elicit useful information. It is more productive and less threatening, as in seeking a history of mobility disorder, to attempt to relate the continence history to significant dates in the annual cycle. Ask where they spent Christmas and their birthday in each of the last few years, whether they were away from home on holidays, and how they managed with the toilet.

The following points are explored on direct questioning.

- *Onset*: recent or remote; whether or not related to an acute illness.
- *Frequency of occurrence*: occasional, daily, several times a day, always.
- *Times of occurrence*: night only, night and day.
- *Quantity*: a little at a time (dribble), a lot (gush).
- *Circumstances*: while asleep, on the way to the toilet, after going to the toilet, any time.
- *Other disturbances*: slowness of onset of micturition, weak stream, frequency, urgency, precipitancy.
- *Awareness*: unaware, feels it coming but no time.
- *Pain*: before, during, or after micturition.
- *Precipitating events*: on standing up, getting out of bed, coughing, yawning, stretching.
- *Associated events*: constipation, diuretics, diabetes, hypnotics, previous catheterization.
- *Consequences*: anxiety, shame, fear, soreness.

From the answers to these questions alone it is possible to gain an impression of whether the patient suffers from:

- outflow obstruction;
- sphincter incompetence;
- unstable bladder;
- retention with overflow.

or any combination of these.

Physical examination

The physical examination aims to detect:

- intrinsic factors;
- extrinsic factors which affect bladder and bowel performance;
- unrelated disease.

The examination therefore includes the following.

1. Examine the abdomen to assess the strength of abdominal muscles, to feel for a full bladder and for faecal masses.
2. Insert a finger in the rectum to feel for faeces and for the prostate gland.

3. Look at the perineum to assess the muscles and soft tissues and the mouth of the female urethra.
4. Inspect the vaginal mucosa for evidence of atrophic vaginitis.
5. Seek evidence of spinal cord disease.
6. Test the urine for sugar, blood, pus, and organisms.
7. In cases of suspected outflow obstruction determine the urine flow rate.

When the cause of the incontinence is not apparent, and when active treatment is contemplated, help in establishing the mechanism can be obtained from urodynamic studies.

Some clinical presentations of incontinence

Four common clinical presentations of incontinence among elderly patients illustrate possible mechanisms, as follows.

1. *The wet bed.* This is often the first manifestation of incontinence of urine. The high nocturnal fluid outflow causes bladder contractions which do not signal a desire to micturate, and fail to arouse the patient. Patients may deny that they wet the bed, but agree that they woke up to find the bed wet.

2. *The puddles on the floor.* The patient feels the sensation to micturate, but has limited time to respond before involuntary micturition occurs. Slowness in rising from bed or chair, walking to the toilet, or transferring to the commode and preparing the clothing uses up the available time. A series of puddles on the trail to the toilet indicates the points at which uninhibited bladder contractions overcame the outflow resistance.

3. *'He is soaking'.* Bladder contractions of great intensity may be induced by posture change or sudden increase in intra-abdominal pressure from some other cause; so that when the patient rises from a chair or yawns the bladder involuntarily empties, and a large lake forms under the chair.

4. '*Every 10 minutes*'. Outflow obstruction, or reduced detrusor tone, prevent the bladder from emptying. Weak detrusor contractions push a small quantity of urine past the obstructed outflow. A few minutes later a further spurt of urine occurs. The patient constantly goes to the toilet and returns dissatisfied and tries again.

Management of incontinence of urine

Management in hospital

Large differences are found in the prevalence of incontinence among similar patients in different environments, as a consequence of difference in the total regime rather than in any one treatment. Severely and persistently incontinent patients are rarely cured; but those who are finely balanced between continence and incontinence are much affected by the regimen. The main elements of a good multidisciplinary regimen in a hospital are as follows.

- There is a team approach, with contributions from doctor, nurse, physiotherapist, occupational therapist and social worker.
- Attainable objectives are set at a weekly case conference and are agreed by all staff.
- Progress towards attainment of the objectives is monitored by effective charting.
- The patient keeps his or her own chart.
- The patient is encouraged to drink freely, especially in the afternoon and early evening.
- Attention is paid to regular bowel movement.
- Sedative and hypnotic drugs are avoided.
- Diuretics are avoided if possible, or mild, long-acting drugs are used.
- Diabetes is controlled.
- Urinary tract infections are treated.
- Pelvic floor exercises are attempted in suitable women.

The hospital environment

A urinal or bedpan is kept within reach of the patient to allay anxiety, but use is made of the normal toilet. This is warm, quiet, private, clean, and within easy reach. There is a hand-rail to help patients manage without the assistance of staff. Chairs are of a type that patients can easily get out of. There is an active programme of physical therapy to keep the patient spry and alert. The patient is trained to dress and undress quickly. Any 'accident' is dealt with quickly and without fuss. Odour control is meticulously maintained by removing and soaking clothing, bedding, and floor coverings as soon as they are soiled.

The night staff play a large part in continence management but do not usually attend case conferences. Their contribution should form part of the assessment and should be included in the management plan.

The construction, equipment, and staffing patterns of many hospitals and nursing homes do not allow the implementation of many of these recommendations. Continence policies require allocation of resources for their implementation; and staff who are dedicated to the reduction of incontinence in their units are entitled to full support from the administration.

Toilet training

Toilet training aims to improve performance by restoring self-control. This requires encouraging the recognition of first desire to micturate, and allowing access to the toilet, while urging the patient to refrain from micturition for a little while and to endeavour to suppress the urge. The hope is that this procedure will restore involuntary inhibition of spontaneous bladder contractions; and will allow a progressive lengthening of the interval between successive acts of micturition.

In busy geriatric wards, where many patients have impaired mobility and cannot reach the toilet in time, toilet training is difficult to achieve. The procedure is often modified by offering toilet facilities every hour initially, and gradually lengthening the interval to two or three hours.

Management at home

The principles of home management are the same as those for hospital care, but their application is more difficult, largely because of the close personal relationship between patient and carer. It may be advisable first to admit the patient to hospital to assess the factors which are causing incontinence, and to establish a continence regimen which can be maintained at home.

Drug treatment

Many drugs have been proposed for the treatment of incontinence. Their effectiveness has been demonstrated as part of a total regimen in selected cases, but their use is no substitute for paying attention to the other factors which contribute to incontinence.

Outcome

The best results in management are obtained when the patient is poised on the balance between continence and incontinence. A minor change in patient or environment can lead to a major change in function.

Management of persistently incontinent patients

When the hopes of cure are low the patient can be made more comfortable and the environment can be protected by:

- body-worn absorbent pads;
- bed pads;
- body-worn devices;
- catheters.

These deal with the environmental consequences of incontinence, not with incontinence itself. They should be neither the first line nor the last resort in treatment.

Body-worn pads

Pads have three layers:

- a cover-stock worn next to the skin which allows the passage of urine in one direction only;
- an absorbent layer;
- a water-resistant backing which prevents contact between the absorbent area and the clothing.

The amount of urine that a pad can absorb depends on its size and the nature of the absorbent material, which are predictable; and on the rate and direction of flow of the urine when the pad is worn, which is not predictable. When a pad has absorbed the maximum of which it is capable, more urine will lie in a puddle on the surface of the pad and irritate the skin; or the pad will burst and the urine will escape into the clothing. As the pad fills it becomes heavier at a rate of approximately 100 gm an hour. After three hours, with normal rates of urine flow, it weighs 300 gm more than it does when dry.

Body-worn pads are helpful in intelligent, ambulant patients with dribbling or occasional incontinence who can change themselves or be changed quickly and frequently; and in some anxious patients with very occasional incontinence who are reassured by the protection which they offer. Their use is inappropriate in patients who pass urine voluntarily into the pad, rather than make the effort to use the toilet.

Bed pads

Early designs of bed pad, many of which are still in use because of their cheapness, are made of layers of cellulose with a paper cover and plastic backing. They have many disadvantages. They disintegrate when overfilled, leaving pieces of soggy paper adhering to the patient's skin. They slip around the bed and may end up over the back or down at the feet. Often several are to be seen in one bed, packed on top of one another, in a vain attempt to increase absorption.

Bed sheets

Absorbent, reusable, textile draw sheets or full-bed sheets are more effective than bed pads, but are expensive and require to be washed and dried in machines.

Body-worn collecting devices

Men may use a pubic pressure bag or a condom drain. There are no suitable body-worn collecting devices for women.

Pubic pressure bags are helpful in well-motivated, ambulant patients but it is difficult to ensure that they do not leak when the wearer sits in a low seat, such as in a motor car, or lies down. The urine bathes the scrotal skin when the wearer sits, and causes irritation.

Condom drains can be successful in the right patient and when well applied. However, because of variation in size of the penis throughout the day and night they may fall off or be pulled off. They are unsuitable for patients with high urine flow-rate, e.g. after a large, uninhibited contracture of the bladder, as this may force the sheath off the penis. They may cause urinary infection.

Catheters

The decision to catheterize a patient should be made by doctor and nurse together, in consultation with the patient, relatives, and community staff who may be involved with further management. The catheter solves some problems but may cause others, notably infection, allergic reactions, leakage, and blockage. Catheters are aesthetically undesirable, and interfere with sexual activity. Careful choice of catheter and a good catheter-management policy reduce these disadvantages.

Leakage

The urethra is shaped like the mouth. When a drinking straw is inserted in the mouth, the lips close in on it, making an air-tight seal, and allowing fluid to be drawn in without spillage. When a wide-bore rubber tube is inserted the mouth gapes open. The lips are unable to grip it, suction is difficult to generate, and fluid spills out of the corners of the mouth. A small catheter in the bladder is gripped tightly by the urethra; a large one is not, and fluid leaks out at the sides.

Leakage due to bypassing is managed by removing the catheter and replacing it with a smaller one.

Blockage

Blockage by debris and gravel can be minimized by choosing a silicon-coated catheter, maintaining a high fluid intake, and preventing constipation. Opinion is divided on the value of bladder wash-outs.

Collecting bags

Collecting bags come in two sizes only, 300 ml and 2000 ml; 2000 ml of urine weighs 2 kg, and 2000-ml collecting bags are intended to be used in bed only. Care has to be taken to avoid compression of the tubing by the patient. Three hundred ml of urine can be collected in an hour or two after drinking one or two cups of tea; 300-ml bags are intended for daytime use by ambulant patients, and need to be emptied frequently.

Tubing

External collecting devices and catheters should be connected by a short length of tubing to a collecting bag worn under the clothing over the thigh. The bag is emptied by a clip at its lower end which allows the urine to escape without breaking the intended sterility of the closed system, although it is arguable whether any urine-collecting system is really 'closed'. Before an external device or catheter is fitted the ability of the patient or carer to manage the collecting bag and tubing and to empty and change the bags should be determined.

Aesthetics

Some patients are managed with a large bag and a long tube. The bag is attached around the knee or calf, but may slip down as it fills, and may protrude from underneath the clothing, or trail along the floor. Sometimes the bag is not attached to the leg, but is carried by the patient.

The urine bag need not and should not be visible to others. A small bag with a short tube can be attached to the thigh, and concealed under the clothing, where it is easily reached by the patient for emptying. Special underpants are available with a pouch for the bag to sit in snugly and modestly. The incontinent patient is entitled to the same privacy as is accorded to an ileostomy patient.

Incontinence of faeces

The odour, texture, and appearance of leaked faeces arouse even stronger feelings of distaste than do those of leaked urine; but these need not adversely affect the help that is given to patients.

Causes

Disturbances of defaecation in elderly hospital patients are due to:

- pathological and age-related changes in bowel function;
- weakness of abdominal musculature;
- changes in consistency of faeces;
- environmental factors.

Changes in bowel function

In elderly, ill people, and especially in the presence of brain failure, changes occur in the rhythmicity of the bowel, the rate of intestinal emptying, the competence of the closure mechanism of the rectum and anus, and the responsiveness of the rectal wall to distension by faeces. These changes are analogous to the changes in bladder function which occur in similar patients.

Massive retention of faeces may occur in chronic illness and in depressive states. This may be accompanied by faecal smearing. Faecal masses are palpable through the abdominal wall. Laxatives and high-fibre diets may be ineffective; and bowel clearance may require days of enemas and manual evacuation.

Abdominal musculature

Contraction of the muscles of the abdominal wall aids defaecation. These muscles are weak in ill, old people.

Changes in faeces

The alteration of diet which occurs on admission to hospital, and the limitation in access to fluids, tend to produce hard, dry stools. Bowel infection causes wet stools. Both changes adversely affect bowel function.

Environment

Hospital toilets are often cold and lack privacy. Toileting regimens may not coincide with times of maximum intestinal motility. The lavatory seat may be at the wrong height for the patient. The use of a commode may be unattractive. These factors tend to cause constipation. Treatment by aperients, suppositories, and enemas may not prevent faecal impaction.

Mechanisms of faecal incontinence

Incontinence of faeces may be due to:

- the unavoidable passage of wet, unformed stools in infective diarrhoea and in pathological states such as ulcerative colitis and rectal carcinoma;
- the bypassing of faecal plugs in hard impaction with oozing of a semiliquid sludge;
- the inadvertent passage of semiformed stools.

Management

Faecal incontinence is preventable by good attention to food and fluid intake, encouragement of normal bowel habit, and correction of environmental disadvantage.

Faecal impaction in the absence of brain failure can be corrected by repeated enemas, suppositories and manual evacuation; and sometimes by aperients and a high-fibre diet, although these may cause diarrhoea and faecal blockage, respectively, and need care in use.

Persistent faecal incontinence in the absence of local bowel disease is usually accompanied by severe brain failure and is very difficult to correct. Success has been achieved by combining the administration of constipating drugs like codeine with twice-weekly enemas.

Conclusion

The understanding of urinary and faecal incontinence and their sympathetic management require knowledge of physiology, pharmacology, pathology, psychology, rehabilitation, group dynamics, the design and function of equipment, and the motivation of staff. These are the skills that go to the making of a complete physician.

Incontinence does not deserve its unpopularity. A geriatric ward in which urine and faeces are neither seen nor smelt is testimony to good management by a multidisciplinary team of a multifaceted problem.

Notes and references

Urinary incontinence

The subject was summarized by Williams and Panvill (1982).

Terminology

The International Continence Society (1976, 1977, 1980, 1981) aimed to standardize the pre-existing confusion of terminology. They introduced the term 'lower urinary tract dysfunction' to cover the symptoms of urgency, frequency, and obstruction, whether or not these were accompanied by incontinence. Detrusor and urethral function were described as normal, overactive, or underactive; and bladder sensation as hyposensitive, normal, or hypersensitive. The term 'low compliance bladder' replaces the 'unstable bladder', in which there is a high pressure rise with small increase in volume. The term 'high compliance bladder' describes the situation in which there is a low rise of pressure with a large bladder volume.

Classification of incontinence

Early classifications, which were based on pathophysiology, urodynamics, clinical observation, or a mixture of these, employed a variety of terms. Resnick and Yallas (1985) reduced these to urge, reflex, stress, overflow, and mixed. Panvill (1987) preferred six categories: sphincter weakness, urethral incompetence, detrusor instability with hyperactivity and urge, overflow, and 'functional'. Resnick (1990) reduced these to hyperactivity, detrusor underactivity, stress incontinence, and outlet obstruction; while Brocklehurst (1990), in a practical field study, was satisfied with urge, stress, and 'unaware until wet'.

Aetiology of incontinence

Factors in addition to urinary tract dysfunction which contribute to incontinence include the passage of large amounts of urine by night (Kirkland *et al.* 1983) and when lying in bed (Guite *et al.* 1988); impairment of mobility, functional ability, toileting skills, and cognition (Isaacs and Walkey 1964; Tobin and Brocklehurst 1986*a, b*; Ouslander *et al.* 1987*b*); diabetes; and the use of diuretic sympathomimetic and sedative drugs (Abrams *et al.* 1983). The factors associated with overflow incontinence include prostatic hypertrophy, bladder-neck obstruction, urethral stricture, underactive detrusor, peripheral neuropathy, diabetes mellitus, tabes dorsalis, and herniated lumbar disc (Resnick and Yallas 1985).

Pathophysiology

The commonest urodynamic findings in elderly nursing-home patients were overactive bladder and outlet obstruction, alone or in combination (Resnick *et al.* 1989). One-quarter of younger patients with stress incontinence also had detrusor instability (McGuire *et al.* 1980).

Overflow incontinence

Resnick (1990) described the bladder as 'an unreliable witness' in the diagnosis of overflow incontinence, since measurement of residual volume is very inconstant (Panvill 1987). It is possible to measure residual volume more accurately by ultrasonography (Brocklehurst 1990). Abrams *et al.* (1983) believed that clinical assessment, together with a low flow rate, selects the majority of patients who have outflow obstruction; Brocklehurst (1990) and Resnick (1990) were satisfied that in the great majority of cases a diagnosis can be made from symptoms and signs alone, without urodynamic studies.

Detrusor hyperactivity

Abrams *et al.* (1983) confirmed the finding of Brocklehurst and Dillane (1966*a, b*) that, among people over the age of 65 attending a urodynamic clinic, an unstable bladder occurred just as often in the continent as in the incontinent. Detrusor instability was common in continent men suffering from urgency and awaiting prostatic surgery (Speakman *et al.* 1987). In elderly, long-term hospital patients, detrusor contractions were poorly co-ordinated and ineffective, the bladder was trabeculated, and there was a high residual volume (Resnick and Yallas 1987).

Urodynamic studies

Urodynamic studies are objective, have stimulated research and have extended knowledge (Abrams *et al.* 1983). Their use in elderly people is recommended in the investigation of reflex and overflow incontinence, when the bladder output is less than 200 ml and the flow rate is reliably low; when surgery is being considered (Resnick and Yallas 1987); and when surgery has failed to correct the incontinence. However, there is a poor relationship between clinical and urodynamic findings (Diokino *et al.* 1982; Abrams *et al.* 1983; Panvill 1987). Among elderly, long-term, institutionalized patients these studies are 'expensive, invasive, unfeasible and unavailable' (Resnick and Yallas 1987); show a poor return for invested effort (Abrams *et al.* 1983); and only rarely do the results influence management (Tobin and Brocklehurst 1986*a, b*).

Cystometry failed in one-third of elderly, incontinent women because of extrusion of the pressure line. A single measurement of flow rate is unreliable, unless it exceeds 12 ms/s. when significant outflow obstruction can be virtually excluded. However, many elderly patients were unable to pass water for this estimation (Abrams *et al.* 1983). Combined urodynamic and videocystourethography is of value in the investigation of the neuropathic bladder (Abrams *et al.* 1983).

Management

Management was discussed by Mandelstam (1986). A scheme of management based on experience in a continence clinic was presented in algorithmic form by Hilton and Stanton (1981). Abrams *et al.* (1983) recommended short-term hospitalization to introduce ambulant, intelligent patients to charting, bladder training, and peroneal exercises, combined with limited use of parasympathomimetic drugs. Castleden *et al.* (1985) advocated judicious use of drugs in out-patients, in combination with exercises and bladder training. Palmer and McCormick (1989) stressed the value of charting by nurses, provided that the forms were readily understood.

Bladder training

Bladder training consists of instruction in simple bladder physiology; recording by the patient of time and volume of micturition; and adopting a graduated, negative response to the urgent desire to micturate (Frewen 1980). Factors which favour the relief of urgency and frequency by this method include: age under 65; being ambulant, intelligent and preferably continent (Castleden *et al.* 1985; Elder and Stephenson 1980; Frewen 1980; Pengelly and Booth 1980); maximum voided volume in excess of 200 ml; frequency of voiding of less than four times in 12 hours; low bladder pressure and good outflow (Castleden *et al.* 1985); variability in the pattern and in the amount of urinary loss (Abrams *et al.* 1983); and the ability to sense a full bladder and to tell someone about it (McCormick *et al.* 1988). Unfavourable factors are the presence of uninhibited detrusor contractions (Elder and Stephenson 1980); and of needing to micturate before the time of the toilet rounds (Petrilli *et al.* 1988).

Catheterization

Brocklehurst and colleagues (Kennedy and Brocklehurst 1982; Kennedy *et al.* 1983; Roe and Brocklehurst 1987) surveyed the use of indwelling catheters in hospital and community. Management was not standardized. All patients were conscious of the catheter, and most found it uncomfortable or painful, although most became used to it after one year. The catheter adversely affected choice of clothes and sexual life.

Leakage and blockage of catheters were common, and were attributable to the use of too big a catheter. Catheters increase in diameter during use, while balloons tend to deflate (Barnes and Malone-Lee 1986), necessitating premature replacement of the catheter. Pure silicon catheters are more permeable than silicon-coated ones, and tend to deflate more readily. When the balloon fails to deflate and cannot easily be withdrawn, Amin and Amin (1987) warn against trying to burst it, and recommend passing a fine wire along the inflation channel to remove the blockage.

Plastic drainage bags constitute an open drainage system. Infection can enter at the meatus, the junction between the catheter and the drainage bag, and the tap of the drainage bag. Air bubbles carrying bacteria may ascend through the flutter valve (Holliman *et al.* 1987). Infection can be reduced by instilling peroxide into the bag; but in patients with long-term catheters, bladder irrigation is unsuccessful (Gopal and Elliott 1988).

Clamping the catheter before it is removed, and emptying the bladder every three or four hours, does not affect the acquisition of continence (Gross 1990).

Intermittent, non-sterile, self-catheterization was recommended by Lapides *et al.* (1976), and found to be effective in old people by Whitelaw *et al.* (1987). Patients must have adequate dexterity, intellect, and motivation. The method is recommended when there is a large residual volume, not associated with a correctable outflow obstruction. A high proportion of cases become infected (Warren *et al.* 1981). The alternative of 'external catheterization' or condom drainage, is also associated with a high rate of infection (Ouslander *et al.* 1987*a*).

Faecal incontinence

The subject is reviewed by Brocklehurst (1989). The classification recommended by Tobin and Brocklehurst (1986*a*) employs a mixture of clinical and pathophysiological terms:

- 'neurogenic', in which a formed stool is passed once or twice daily, and the rectum is empty;
- 'secondary to faecal impaction', with continuous faecal soiling and a lax anal sphincter;
- 'soiling without faecal impaction', which is a behavioural disorder;
- 'colorectal pathology' associated with diarrhoea.

Aetiology

Constipation and delayed intestinal transit time accompany the first two types, and probably the third (Melkersson *et al.* 1983). Associated factors are: low-fibre diet; lack of physical exercise; abuse of laxatives; neglect of the physiological call to stool on rising or after the first food of the day (Castle 1989); and denervation of the anal sphincter with increase of fibre density (Percy *et al.* 1982).

Management

Tobin and Brocklehurst (1986) recommend treating constipation and faecal impaction by daily enemas until the rectum is empty, then giving Lactulose twice daily, and an enema once weekly. Patients whose stools are formed or loose are given codeine phosphate daily and a weekly enema. Puxty and Fox (1986) advocate clearing faecal impaction with a rectal infusion of polyethylene glycol.

References

Abrams, P., Feneley, R., and Torrens, M. (1983). *Urodynamics*. Springer Verlag, New York.

Amin, E. D. and Amin, M. (1987). Management of non deflatable Foley balloon catheter. *Journal of the American Geriatrics Society*, **35**, 886–7.

Barnes, K. E. and Malone-Lee, J. (1986). Long term catheter management: minimizing the problem of premature replacement due to balloon deflation. *Journal of Advanced Nursing*, **11**, 303–7.

Brocklehurst, J. C. (1989). The problem of faecal incontinence. In *Gastrointestinal tract disorders of the elderly* (ed. G. Hellemans and G. Vantrappen) Churchill Livingstone, Edinburgh.

Brocklehurst, J. C. (1990). Urinary incontinence in old age: helping the general practitioner to make a diagnosis. *Gerontology*, **36** (Suppl. 2), 3–7.

Brocklehurst, J. C. and Dillane, J. B. (1966*a*). Studies of the female bladder in old age. I. Cystometrograms in non incontinent women. *Gerontologia Clinica*, **8**, 285–305.

Brocklehurst, J. C. and Dillane, J. B. (1966*b*). Studies of the female bladder in old age. II. Cystometrograms in 100 incontinent women. *Gerontologia Clinica*, **8**, 306–19.

Castle, S. C. (1989). Constipation: endemic in the elderly? Gerontopathophysiology, evaluation and management. *Medical Clinics of North America*, **73**, 1497–509.

Castleden, C. M., Duffin, H. M., Asher, M. J., and Yeomanson, C. W. (1985). Factors influencing outcome in elderly patients with urinary incontinence and detrusor instability. *Age and Ageing*, **14**, 305–7.

Diokino, A. C., Wells, T. J., and Brink, C. A. (1982). Urinary incontinence in elderly women: urodynamic evaluation. *Journal of the American Geriatrics Society*, **35**, 940–6.

Elder, D. D. and Stephenson, T. P. (1980). An assessment of the Frewen regime in the treatment of detrusor dysfunction in females. *British Journal of Urology*, **52**, 467–71.

Frewen, W. K. (1980). The management of urgency and frequency of micturition. *British Journal of Urology*, **52**, 367–9.

Gopal, Rao, G. and Elliott, T. S. J. (1988). Bladder irrigation. *Age and Ageing*, **17**, 373–8.

Gross, J. C. (1990). Bladder dysfunction after a stroke: it's not always inevitable. *Journal of Gerontological Nursing*, **16**, 20–5.

Guite, H. F., Bliss, M. R., Mainwairing-Burton, R. W., Thomas, J. M., and Drury, P. L. (1988). Hypothesis: posture is one of the determinants of the circadian rhythm of urine flow and electrolyte excretion in elderly female patients. *Age and Ageing*, **17**, 241–8

Hilton, P. and Stanton, S. L. (1981). Algorithmic method for assessing urinary incontinence in women. *British Medical Journal*, **282**, 940–2.

Holliman, R., Seal, D. V., Archer, H., and Doman, S. (1987). Controlled trial of chemical disinfection of urinary drainage bags. *British Journal of Urology*, **60**, 419–22.

International Continence Society (1976, 1977, 1980, 1981). Reports on the standardisation of lower urinary tract function. *British Journal of Urology*, **48**, 39–42; **49**, 202–10; **52**, 348–50; **53**, 333–5.

Isaacs, B. and Walkey, F. A. (1964). A survey of incontinence in the elderly. *Gerontologia Clinica*, **6**, 367–76.

Kennedy, A. P. and Brocklehurst, J. C. (1982). The nursing management of patients with long term indwelling catheters. *Journal of Advanced Nursing*, **7**, 411–17.

Kirkland, J. L., Lye, M., Leny, D. W., and Bannerjee, A. K. (1983). Patterns of urine flow and electrolyte excretion in healthy elderly people. *British Medical Journal*, **287**, 1665–7.

Lapides, J., Diokino, A., Goulf, F. R., and Lowe, B. S. (1976). Further observations on self catheterization. *Journal of Urology*, **116**, 169–71.

McCormick, K. A., Schele, A. A. S., and Leahy, E. (1988). Nursing management of urinary incontinence in geriatric inpatients. *Nursing Clinics of North America*, **23**, 231–64.

McGuire, E. J., Lytton, B., Kohorn, E. I., and Pepe, V. (1980). The value of urodynamic testing in stress urinary incontinence. *Journal of Urology*, **124**, 256–8.

Mandelstam, D. (1986). *Incontinence and its management*. Croom Helm, Beckenham, Kent.

Melkersson, M., Andersson, H., and Bosaeus, I. (1983). Intestinal transit time in constipated and non-constipated geriatric patients. *Scandinavian Journal of Gastroenterology*, **18**, 593–7.

Ouslander, J. G., Greenwood, B., and Chen, S. (1987*a*). External catheter use: urinary tract infection among incontinent male nursing home patients. *Journal of the American Geriatrics Society*, **35**, 1063–70.

Ouslander, J. G., Morishita, L., Blaustein, J., Orzeck, S., Dunn, S., and Sayre (1987*b*). Clinical, functional and psychosocial characteristics of an incontinent nursing home population. *Journal of the American Geriatrics Society*, **42**, 631–7.

Palmer, M. H. and McCormick, K. A. (1989). Do nurses consistently document incontinence? *Journal of Gerontological Nursing*, **15**, 11–16.

Panvill, F. C. (1987). Urinary incontinence. *Journal of the American Geriatrics Society*, **35**, 880–2.

Pengelly, A. W. and Booth, C. M. (1980). A prospective trial of bladder training as treatment for detrusor instability. *British Journal of Urology*, **52**, 463–6.

Percy, J. P., Neill, M. E., Kandian, T. K., and Swash, M. (1982). A neurogenic factor in faecal incontinence in the elderly. *Age and Ageing*, **11**, 175–9.

Petrilli, C. O., Traughber, B., and Schnelle, J. F. (1988). Behavioural management in the inpatient geriatric population. *Nursing Clinics of North America*, **23**, 265–77.

Puxty, J. A. H. and Fox, R. A. (1986). Golytely: a new approach to faecal impaction in old age. *Age and Ageing*, **15**, 182–4.

Resnick, N. M. (1990). Noninvasive diagnosis of the patient with complex incontinence. *Gerontology*, **36** (Suppl. 2), 8–18.

Resnick, N. M. and Yallas, V. (1985). Management of urinary incontinence in the elderly. *New England Journal of Medicine*, **313**, 800–5.

Resnick, N. M. and Yallas, V. (1987). Detrusor hyperactivity with impaired contractile function: an unrecognised but common cause of incontinence in elderly patients. *Journal of the American Medical Association*, **257**, 3076–81.

Resnick, N. M., Yallas, V., and Laurino, E. (1980). The pathophysiology of urinary incontinence among institutionalised elderly persons. *New England Journal of Medicine*, **320**, 1–7.

Roe, B. H. and Brocklehurst, J. C. (1987). Study of patients with indwelling catheters. *Journal of Advanced Nursing*, **12**, 713–18.

Saire, J. (1987). Clinical, functional and psychosocial characteristics of an incontinent nursing home population. *Journal of Gerontology*, **42**, 631–7.

Speakman, M. J., Sethia, K. K., Fellows, G. J., and Smith, J. C. (1987). A study of the pathogenesis, urodynamic assessment and outcome of detrusor instability associated with bladder outflow obstruction. *British Journal of Urology*, **59**, 40–4.

Tobin, G. W. and Brocklehurst, J. C. (1986*a*). Faecal incontinence in residential homes for the elderly: prevalence, aetiology and management. *Age and Ageing*, **15**, 41–6.

Tobin, G. W. and Brocklehurst, J. C. (1986*b*). The management of urinary incontinence in Local Authority residential homes for the elderly. *Age and Ageing*, **15**, 292–8.

Warren, J. H., Muncie, H. L., Bergquist, E. J., and Hoopes, J. M. (1981). Sequelae and management of urinary infection in the patient requiring chronic catheterization. *Journal of Urology*, **125**, 1–8.

Williams, M. E. and Pannill, F. C. (1982). Urinary incontinence in the elderly: physiology, pathophysiology, diagnosis and treatment. *Annals of Internal Medicine*, **72**, 895–902.

Aphorisms

Of attitudes

- Most people have wet the bed or soiled their clothes at some stage of their life, but prefer to forget about it.
- Close contact with human excreta is not commonly a source of pleasure.
- The management of incontinence is dependent more on the attitude of doctors than on their knowledge or skill.
- Attitudes to incontinence are compounded of antipathy, apathy, sympathy, and empathy.
- Antipathy is disgust; apathy is distance; sympathy is care; empathy is action.
- The antipathetic scolds; the apathetic catheterizes; the sympathetic is sorry; the empathetic investigates and treats.

Of cause and effect

- Behind every full bladder there lies a full rectum.
- Defaecation is the privilege of the independent. Impaction is the fate of the dependent.
- Patients respond to frequency of micturition by infrequency of imbibition.
- The incontinent absolve but do not deny. To the question 'Did you wet the bed?' they reply 'No'; to the question. 'Did you find the bed wet?' they reply, 'Yes.
- Incontinence is involuntary; para-incontinence is erroneous. Incontinence is failure of control; para-incontinence is failure of comprehension.

Of management

- Ignorance of the rules of incontinence is as inexcusable as ignorance of the rules of cricket.

- Conversations between doctors and patients on the subject of bladder and bowel function are curious. The patient uses the doctor's words but does not understand them; the doctor understands the patient's words but does not use them.
- In the management of incontinence a few can be cured, many can be improved and all can be better understood.
- The smearing of faeces on the wall is not easily seen as a plea for help.
- Toilet training is like Marilyn Monroe, well regarded but little understood.

Of pads and catheters

- Catheterization of the bladder, like the sacrament of matrimony, is not to be entered upon lightly or wantonly.
- Leaking round a catheter means that the catheter is too big, not that it is too small.
- A small catheter in the urethra is like a drinking straw in the mouth; a larger catheter in the urethra is like a rubber tube in the mouth. You do not leak round a drinking straw but you do round a rubber tube.
- Incontinence pads are good for dribblers, but bad for gushers.

10 *Intellectual impairment 1: the concept of brain failure*

Introduction

Most people observe, as they approach old age, some diminution in memory and concentration; and say half jokingly that they are becoming demented. The great majority do not progress beyond this state of 'benign senescent forgetfulness'. This may be normal; or it may be the effect of minimal pathological change in the brain. It may cause anxiety and distress, and impair work performance in those approaching retirement. But its significance is mild compared with the damage done by the Giant, Intellectual Impairment.

Terminology

The term 'intellectual impairment' is used as an aide memoire for the Giants, but does not sufficiently describe the clinical syndrome caused by diffuse disease of the brain. Nor do many other terms in common use, such as senility, confusion, altered mental state, cognitive impairment, and chronic brain syndrome.

The terms 'dementia' and 'delirium' have gained the official sanction of the American Psychiatric Association. There is, however, a need for a general term which can be understood by non-psychiatrists and lay persons. For this purpose the term 'brain failure' is proposed.

Definition of brain failure

Brain failure is a clinical syndrome in which the brain as a whole functions abnormally, as a result of diffuse or widespread pathological change in the brain tissue and/or functional change in its nutrition. The term thus includes both 'dementia' and 'delirium'; but excludes the clinical syndromes which result from discrete focal lesions of the brain, such as haemorrhage, infarct, tumour, and abscess. Brain failure implies malfunction of all or nearly all of the brain.

The term 'brain failure' is analogous to 'heart failure', which also describes a clinical syndrome with many causes. Brain failure is, however, less uniform than heart failure, more complex, and more individual; because the brain mediates the infinite complexities of thought and behaviour; and contains a record of the unique life of its owner.

Causes of brain failure

The causes of brain failure are summarized in Fig. 10.1. Intrinsic brain failure tends to develop slowly and to run a long, progressive course. Extrinsic brain failure tends to develop rapidly and to run a short, reversible course. Combined extrinsic and intrinsic brain failure tends to cause an acute exacerbation followed by return to the former level or to a lesser level of brain function.

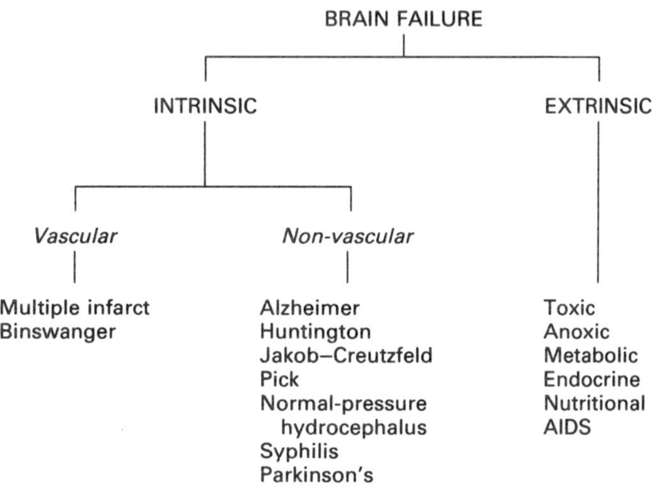

Fig. 10.1 Cases of brain failure.

The mechanisms of brain failure

The features of brain failure include wandering, confusion, disorientation for time, place, and person and incontinence. The patient with brain failure has retained some brain functions and lost or altered others.

The features which are retained include:

- logical thought;
- intelligence;
- personality;
- stored memory and skills;
- emotion.

The lost and altered features are:

- loss of registration and recall;
- perceptual error;
- dissociation between percept and concept;
- dissociation between cause and effect;
- impairment of time–space coordinates.

The normal features

Logical thought

Patients with brain failure think and act logically, but from wrong assumptions. For example, every time a patient heard a nearby factor hooter sounding, he left the ward in order to go home. He retained the logical association between the hooter and time to finish work; but wrongly assumed that he was living 20 years ago and was still at work.

Intelligence

Patients retain their reasoning power and can be surprisingly sharp and astute. Under pressure from questioning they show craft and ingenuity in avoiding a direct answer. For example, a patient who was asked to name the prime minister, in the reign of Mrs Thatcher, replied 'I can see that you are trying to make a fool of me, so I am not going to tell you, but I know perfectly well who he is.'

Personality

The uniqueness and strength of individual personality are retained late into the disease. An example of this was a retired University Professor, who organized his life as a victim of brain failure by placing, at all points of danger or need in his house, index cards bearing reminders of necessary domestic tasks.

Stored memory and skills

Patients retain vivid memories of themselves as they were. They maintain acquired skills like playing the piano and the niceties of social conversation, and they retain lifelong habits. An example was a Scottish patient with advanced brain failure who celebrated Burns Night by reciting the whole of 'Tam o' Shanter' word perfect with every intonation and gesture.

Emotion

The range and intensity of emotion persists into the advanced stages of brain failure. Patients suffer anxiety, fear, resentment, depression, jealousy, and rage; as well as offering love, affection, concern, sympathy, and unselfishness. For example, a man of 86 who had been married for 65 years followed his wife everywhere, even into the toilet, and could not bear to be parted from her.

Abnormal features

Loss of registration and recall

Recent events and new experiences, especially those which cannot be fitted into an existing schema, are not registered and therefore they are not recalled. The following story is typical.

> A woman 'drove her daughter mad' because she kept on asking the same question, no matter how often it was answered. She continually lost her handbag, keys, pension book, and money. She was in the habit of phoning her daughter every few minutes to ask her to come over to help her to find what was missing, and indignantly rejected her daughter's complaint that she had already phoned about the same subject 10 times, that the daughter had come over to her house and found the handbag for her, and had only that minute returned home.

Perceptual error

Perceptions in brain failure are often erroneous. Normal people see dissimilarities between similar objects; patients with brain failure see similarities between dissimilar objects. The resultant behaviour is perceived as abnormal or antisocial. In advanced brain failure, objects which belong to the same 'set' are not distinguished from one another. The patient may see no difference between passing urine into a toilet bowl and passing it into a flower bowl or a cooking bowl or a kitchen sink, all of which belong to the set of objects which hold water. He may see no difference between micturating against the wall of a urinal and against the wall of the ward.

Dissociation between percept and concept

A percept is what we perceive; a concept is what we believe. Concepts include beliefs about persons, place, and time.

When normal persons perceive a woman working in a hospital who is wearing distinctive clothing, they match this percept to their previous knowledge and experience; and they arrive at two concepts: that the person is a nurse, and that her clothing is a uniform. Patients with brain failure do not derive concepts from percepts, because they lack access to information stores of past experience. A patient who was shown a nurse and was asked 'Who is that?' said, 'A woman'; and to the question 'What is she wearing?' he replied, 'A nice dress.'

When a normal person encounters the unexpected, or in other words, when a percept does not match a concept, another area of experience has to be tapped in order to create a new concept which resolves the uncertainty. For example, in a psychiatric ward, a person may be seen caring for a patient but wearing normal clothing. She is identified as a nurse by invoking the concept that nurses in psychiatric wards do not wear uniform. However, even normal people may retain a false concept despite contradictory percepts; for example they may believe that the clocks are fast rather than that their watch is slow.

Patients with brain failure are unable to resolve a mismatch between percept and concept, but instead they retain their original concept and ignore percepts which do not fit; or else they invent 'explanations'. A patient formed the concept that the nurse was her daughter. When her daughter came to visit her, she still insisted on calling the nurse her daughter, but the visitor was 'her other daughter'.

Disorientation for time, place, and person in patients with brain failure results from the formation of a concept based on a single percept, which is retained despite contrary evidence. An example was a patient who saw a budgerigar in the ward. He had a similar budgerigar in his own home, and he formed the concept that he was at home. When he was asked to explain the presence of other people and beds in the ward he said, 'They must have come to see the budgerigar.' Similarly the patient who left the ward when the factory siren blew described the other patients as fellow workers.

Dissociation between cause and effect

Patients with brain failure do not perceive the link between an event and its consequences. The future does not exist for them. This can cause domestic accidents. For example, an old lady turned on the gas and did not light it. She denied that turning on the gas was responsible for the smell of gas in the room. She boiled a kettle, but denied any connection between that and the kettle whistling, or the emission of steam, or the outpouring of smoke when the kettle boiled dry. She lit cigarettes by igniting a rolled piece of newspaper on an open electric fire and then tossed the burning paper on the floor, but said that that had nothing to do with the burns in the carpet. She left a bath tap turned on for hours but this was not the cause of the subsequent flood.

People with brain failure are often described as thoughtless and irresponsible; but this is because they have lost the ability to perceive the effects of their behaviour.

Impairment of time–space coordinates

Normal persons locate themselves precisely and consistently in time and space. Their here is always here and their now is always now. Patients with brain failure err and fluctuate in their location in time and space. Their here is not always here and their now is not always now.

The logic of a patient's behaviour stems from his or her location in time and space. As brain failure advances in severity, location in space and time recede. The patient believes him/herself to be the person s/he was 10, 20, 30, 40, 50, and more years ago, and to be living as s/he was then in the world as it was then.

An example of this behaviour was seen when a 90-year-old woman packed a case and left the ward. She furiously resisted being brought back by the staff. She insisted that she was on her way to nurse her sick mother in Wales. Her 'here' was not the same as the 'here' of the staff, and her 'now' was not their 'now'. Her behaviour was logical in relation to her orientation, and she perceived the behaviour of the staff as illogical.

Symptoms of brain failure

The main manifestations of the later stages of brain failure are wandering, confusion, disorientation for time, place, and person, incontinence, sleeplessness, disturbances of language, sensory impairment, and violent behaviour.

Wandering

This term implies aimless activity which is socially undesirable or dangerous. The wandering of the patient with brain failure may, however, be purposeful. Wandering can be classified as:

- amnestic;
- anachronistic;
- anxiolytic;
- atavistic.

Amnestic movement is seeking something and forgetting what is being sought. An example was a woman of 84 who had lost a child under rubble in an air raid in the Second World War. She repeatedly made her way back to the street where the bomb had fallen to look for her child.

Anachronistic movement is going somewhere that was appropriate once and no longer is so, like the patient who went home from the ward when the factory siren sounded; or like an 80-year-old retired policeman who returned to his former beat, and arrested innocent passers-by.

Anxiolytic movement is walking to achieve tranquillity. A lady of 90 was frequently seen walking the streets at any time of day or night, oblivious to time of day, weather, or clothing. She had arguments with her family, which ended with her slamming the door and walking until she forgot why she was angry.

'*Atavistic*' wandering has no discernible purpose, and occurs in the advanced stages of brain failure. It is seen in the long corridors of old-fashioned mental hospitals, which virtually encourage this pattern of behaviour.

Confusion

This term may embrace:

- dysphasia;
- delusions;
- hallucinations;
- mood disturbances;
- disorientation;
- misidentification;
- perceptual errors.

It is better to avoid the use of the word confusion in favour of an accurate description of the clinical phenomena.

Disorientation

Correct orientation comprises accurate perception of the here and now; and placement of the percept in an appropriate concept or frame of reference. Disorientation is mismatch between percept and concept.

Errors of percept may occur in brain failure because objects and persons are mistaken for similar but different ones. Errors of concept occur because abstract thinking is impaired. The clinical features of disorientation stem from the patient's bewilderment at the mismatch between percept and concept.

Incontinence

Incontinence of urine and faeces is common in hospitalized patients in the late stages of brain failure, when it represents the combined effects of loss of cerebral inhibition, and exposure to an unfamiliar routine.

Patients at home in the earlier stages of the disease retain the cerebral ability to control micturition and defaecation, but may change their habits in a socially undesirable way, because of errors in perception or conceptualization. They may pass their urine and faeces in inappropriate places, at inappropriate times, and with disregard for social conventions. They may show lack of concern for cleanliness and decency. Urine may be passed into the kitchen sink, flower pots, vases, or cooking utensils. Faeces may be made into little parcels and placed in drawers.

Sleep

The alterations in the normal timing of sleep and wakefulness which characterize brain failure are a major factor in the stress experienced by relatives. Patients may spend much of the day asleep in a chair, while at night they wander restlessly throughout the house or go out ostensibly to work. Sleep is reduced in amount and depth, and is fragmented. As the condition advances the sleep disturbance worsens. The relatives naturally hope that the patient can be made more restful by hypnotic drugs, but these are rarely effective, and in many cases they worsen the situation.

Language

The early changes in language are subtle, and may go unnoticed. There is an impoverishment of language, and a tendency to favour familiar expressions over the creation of new, original sentences. The patient's extensive use of clichés impairs the ability to convey meaning to others. Comprehension of language is reduced, so that long and complex sentences are not understood. There is loss of ability to distinguish between the literal and the metaphorical use of words. Sometimes this difficulty conveys the impression that the patient is joking, as for example, when a psychologist said to a patient, 'I think you see through me', and he replied, 'No I don't, you are not made of glass.' As the condition progresses, sentences lose their grammatical structure and become fragmented; while in the late stage of the disease all ability to speak and to understand may be lost.

Sensory impairment

Patients with brain failure are as liable as others of their age group to suffer from impaired function of the sense organs; but, because of damage to the central connections for vision, sound, taste, smell, and touch, they carry the additional burden of reduced ability to interpret the faulty signal. They are thus liable to misinterpret the sensory environment.

Violence

Violent behaviour in people with brain failure is commonest in those of violent disposition, but even normally mild people may terrify and distress relatives, carers, and other patients in hospital by sudden bouts of violence. This is sometimes apparently unprovoked, but usually results from the patient's terror at the misinterpreted action of someone else. For example a patient lashed out at a nurse who was giving him an injection, but the patient was a former concentration-camp inmate, and believed that he was back in the camp. Violent behaviour is managed by recognizing and correcting its causes, and by ensuring in advance that the patient correctly understands what is going to be done. The unfortunate victims ask for the patient to be sedated, but this rarely helps, and it carries all the risks of obtunding the patient.

Conclusion

The manifestations of brain failure are infinitely complex; but can usually be understood as the behaviour of a bewildered person trying to find his or her way with inaccurate and inadequate information through a world that does not operate according to their rules. These patients can best be helped by attempting to understand why they behave as they do.

11 *Intellectual impairment 2: assessment and history*

Form and aims of assessment

The questions to be answered in the assessment of the patient with suspected brain failure are as follows.

- Is brain failure present?
- What is its duration?
- What are its manifestations?
- What is its cause?
- What are its consequences?
- What else is wrong?

The assessment comprises:

- taking the history from relatives;
- interviewing the patient;
- observing the patient;
- physical examination and diagnostic tests;
- mental testing.

The aims of the assessment are to detect conditions which can be treated, and to give an accurate prognosis. The conditions which are sought may be causal, contributory, consequential, or coincidental. Patients need not be subjected to expensive or uncomfortable tests which merely satisfy curiosity. A computerized scan of the brain is essential in brain failure of recent onset and rapid course, but adds little to the management of a woman of 89 with a 10-year history of progressive brain failure and no neurological abnormalities on clinical examination.

Taking the history from relatives

Patients with brain failure are inaccurate observers and recorders of their own behaviour. If they have brain failure they cannot give an accurate history. If they give an accurate history, they do not have brain failure. The history has to be obtained from those who have observed the progress of the illness.

Relatives and patients are interviewed separately. If the patient is interviewed in the presence of relatives the relatives may answer questions for the patient, interrupt the patient's answers, contradict his or her statements, or visibly indicate their disagreement.

The interview with relatives is often highly emotionally charged, reflecting the tension, anxiety, frustration, guilt, and shame which may have been experienced. It needs time, privacy, and trust. It is largely a matter of listening rather than of probing

Is brain failure present?

The events described by relatives fall into three categories, as follows.

1. Impaired memory and withdrawal from familiar tasks

This may be attributed by the relative to:

- normal ageing;
- strain;
- overwork;
- trauma;
- bereavement;
- retirement;
- impaired vision;
- impaired hearing;
- other physical illness;
- depression.

2. Failure to perform properly tasks which were previously undertaken

This might include:

- neglect of appearance and personal hygiene;
- failure to prepare and eat meals;
- withdrawal from social contacts;
- loss of interest in hobbies and in other people;
- undue irritability;
- misuse of words;
- 'breaking off' in the middle of a sentence;
- impairment of normal sentence structure.

3. Socially disruptive or dangerous behaviour

The following list includes some of the commoner manifestations:

- leaving the gas turned on unlit;
- boiling kettles dry;

- burning saucepans;
- refusing to change clothes;
- refusing to be bathed;
- making excessive phone calls;
- eating voraciously;
- talking about dead people as though they were still alive;
- getting lost in a familiar environment;
- irresponsible use of money;
- making excessive, unecessary purchases;
- unexplained absences;
- 'turning night into day', i.e. doing things at inappropriate times;
- following someone around the house;
- failure to recognize a familiar person;
- misidentifying a familiar person as a stranger or vice versa;
- soiling the bathroom with excreta;
- soiling the clothing or the hands with excreta;
- talking or gesticulating to people who are not there;
- hallucinating;
- exposure or other sexually uninhibited behaviour.

While impaired memory and social withdrawal may be due to depression, brain tumour, or deafness, the socially disruptive behaviour described is characteristic of brain failure.

What is its duration?

In brain failure of extrinsic cause the onset is usually sudden and the course brief. In brain failure of intrinsic cause the onset is usually gradual or insidious and the course prolonged. This is a good rule of thumb, but is not always so, e.g. brain failure due to hypoglycaemia may have a sudden or an insidious onset, and a short or very prolonged course.

The following questions are useful:

- What made you first suspect that something might be wrong?
- When was this?
- Looking back now was there anything which you dismissed at the time but which might have been an indication of something far wrong?
- Were there times when he was his normal self?
- Did you ever say to yourself, That is not my Dad (or Mum or husband or wife)?

What are its manifestations?

In addition to those listed above, relatives may report idiosyncratic features which reflect persistence in distorted form of lifelong habits or peculiarities.

A lady with political interests made luncheon parties to which she invited members of the Cabinet. She set places at table for them, put their photographs on their plates, and chided them for not eating.

A woman of 84 living in a high-rise block of flats caused offence to neighbours by waking them in the middle of the night and offering them cups of tea. This had been one of her duties when she worked as a chambermaid in a hotel.

What is its cause?

The following questions may throw light on causation:

- *Is there a history of brain failure in close relatives?*
  This is sometimes the case in Alzheimer's disease.
- *Has the patient suffered from high blood pressure, heart disease, stroke, or peripheral vascular disease?*
  This raises the possibility of multi-infarct disease.
- *Has he sustained a head injury?*
  This could indicate recent subdural haemorrhage, old subdural haematoma, or subsequent normal-pressure hydrocephalus.
- *Has he sustained recurrent head injuries?*
  In boxers this causes the condition known as dementia pugilistica or 'punch drunkenness' in which repeated traumatic damage is inflicted on the brain.
- *Has he ever had meningitis or a subarachnoid haemorrhage?*
  These may be precursors of normal pressure hydrocephalus.
- *Is there a history of diabetes?*
  Acute hypoglycaemia episodes or chronic hypoglycaemia due to long-acting oral hypoglycaemic drugs may initiate brain failure.
- *Has he received an anaesthetic?*
  Episodes of hypoxia during anaesthesia may induce brain failure.
- *Has he ever suffered cardiac arrest?*
  If the arrest was prolonged, irreversible brain damage may have been caused.
- *What drugs does the patient take?*
  Inappropriate drug therapy commonly causes extrinsic brain failure and aggravates intrinsic brain failure. The drugs which are most often responsible include hypnotics, tranquillizers, sedatives, antihypertensives, antiparkinsonian drugs, and antidepressants.
- *What is his alcohol consumption?*
  High alcohol consumption throughout life sometimes causes brain failure as a result of permanent brain damage. High consumption in late life causes extrinsic brain failure.
- *Has he lost weight?*
  This might point to a malignant lesion causing brain failure due to brain

secondaries or as a metabolic result of the tumour. Weight loss might also be a consequence of malnutrition.

- *Has he been ill in any other way?*
  Brain failure is sometimes a feature of hepatic or renal failure, or is associated with severe anaemia, dehydration, and electrolyte disorders.
- *Has he been depressed?*
  Patients with severe depression may be lethargic and lose contact with reality. This clinical picture has been unsuitably dubbed 'pseudodementia'. The temptation to create new diseases by attaching the prefix 'pseudo' is better resisted. Depression may resemble or coexist with brain failure, but it remains depression and not pseudo anything.

What are its consequences?

These include:

- *Effects on nutrition.* Patients may neglect their food, eat idiosyncratically, or gorge themselves.
- *Compliance with medication.* This is erratic and may aggravate brain failure or cause complications such as falls.
- *Social and interpersonal relationships.* These are seriously disrupted in most cases, and this factor may dominate the interview.
- *Finances.* This is often a major source of concern. The matter should be touched upon to indicate awareness, but details are better dealt with by the social worker.

Sexual behaviour

Relatives may be perplexed and distressed by sexual aberration in a previously respectable person, and may be reluctant to raise the subject, but will respond when told 'It is not unusual for patients with this condition to behave sexually in a way that they would never have done when they were well. Has anything of that nature taken place?'

Mental well-being of carers

Spouses in happy marriages show remarkable tolerance in their handling of a partner with brain failure, although they are very distressed by the 'death' of their loved one while he or she is still alive. Daughters and other carers with a life of their own to live come under severe strain, and it may be appropriate to ask them the startling question, 'Have you ever felt like choking the patient?' This question may bring a sudden look of recognition, relief, and appreciation that someone understands.

What else is wrong?

Relatives are asked for information about other previous or accompanying illnesses which may require further assessment.

'Happenings'

The following are among the 'happenings' which relatives may report:

- *The telephone bill.* The telephone bill rises steeply at the start of the illness because the patient telephones many times a day, to report a loss or to ask a question that has been answered often before. Sometimes the call comes in the middle of the night, because the patient does not recognize the distinction between night and day. The patient does not recall having asked the question and forgets having been answered, but does not forget the anxiety that generated the question.
- *The lost pension book.* The pension book is repeatedly lost but is found in the handbag where it was put. The patient has not registered having put it there, and has not perceived it to be there when searching for it. Someone may be accused of having stolen it.
- *The missing person.* The patient goes out and fails to return for many hours. There is great anxiety, the police are notified, but eventually the patient turns up without an explanation and wonders what all the fuss is about.
- *Locked out.* The patient loses the keys, or goes out without them, or steps outside and the door blows shut. It can happens to anyone, but it happens repeatedly to patients with brain failure.
- *Escaped gas.* The sequence of actions required for making a cup of tea can be broken at any stage by distraction. If it is broken between turning on the gas and lighting the flame, gas escapes. The gas is not smelt; or if it is smelt it is not recognized; or if it is recognized it is not associated in the mind of the patient with having turned on the gas; or with the possibility of a dangerous build up of gas in the room.
- *The burned kettle.* The patient puts on the kettle or pot, but then forgets. The escaping steam, the smoke and the smell of burning, the whistle of the kettle are not clearly perceived; and if they are perceived they are not connected with having put the kettle or pot on to boil. The kettle burns out or the pot boils over.
- *The electric kettle.* After two or three kettles have been burnt out on the gas the relatives provide an electric kettle which switches itself off automatically when it boils. But the patient puts it on the gas, because he or she recognizes the similarities and fails to recognize the dissimilarities.
- *The soiled bathroom floor.* Patients become strangely indifferent to lifelong habits of cleanliness and circumspection. The floor is soiled and they do not care.

- *The bath-day fight*. Bathing and personal hygiene are neglected and then resisted. Patients who were once meticulous now refuse to change their clothes, to wash their hair, to cut their nails, to wipe their bottoms.
- *Dressing*. Old and shabby clothes are worn. Night clothes are worn by day. Summer clothes are worn in winter. In the later stages of the disease the clothes are put on in the wrong order or back to front, and the night attire is put on top of the day clothes.
- *Nudity*. Patients walk about inadequately or inappropriately dressed. Men leave their flies unbuttoned or unzipped, and the penis may hang out. Women expose their breasts or fail to wear knickers. Sometimes they walk indoors or outdoors totally nude.
- *Wasted money*. The patient purchases the weekly shopping as for a full family when the household consists of only one or two persons; and then goes out and purchases the same again. Money is spent or given away without regard to its value.
- *Sexual misbehaviour*. Unpleasant incidents may occur of exposure, nudity, masturbation, suggestiveness, and open sexual advances.
- *'I can't go to the toilet'*. The patient follows the carer around like a shadow and will not leave her, not even when the carer wants to go to the toilet. This behaviour seems to be an act of anchoring to the one sure and recognizable feature of a mysterious, uncomprehended world.
- *'I'm the bad one'*. Dementia coarsens sensitivity and reduces the use of the social niceties. Carers are aggrieved by the loss of familiar courtesies, affection and appreciation. Requests become demands; 'just a minute' becomes 'at once'; the needs of the carer are subservient to the needs of the patient; 'please' and 'thank you' are lost or put to manipulative use. But social behaviour is resumed like a magic cloak the instant anyone else appears on the scene. The carer cannot help resenting that she is treated differently from those who are much less closely involved with the patient.
- *Lability*. Change of mood and of performance in the dement is instantaneous. Fury gives way to charm, incompetence to remembered skills in a second. A new arrival in a room finds the patient a delight, while the carer who has been there all the time knows another side of the story.
- *Talking to the television*. Television characters are perceived as real people who have entered the living room and who are welcomed or resented, feared or ignored.

Summary of interviewing the relatives

The stories that relatives have to tell are full of bizarre and perplexing incidents which tell of the struggles of a patient to maintain contact with a former reality and style of living in the presence of increasing difficulty to remember, to conceptualize, to connect, to identify. The patient eventually loses anchorage to time, place, and person, and drifts in a sea of bewilderment, pursued by the carers. This interview is the major source of information and defines the necessary management.

Interviewing the patient

The patient who is suspected of brain failure may be pleasant and relaxed, or suspicious and aggressive. He or she should not be approached with pad or case notes in hand by a stranger prepared to write down answers to a string of questions, as this increases patients' suspiciousness and makes them recalcitrant and uncommunicative. 'Formal' tests of mental function in which the patient is asked to perform a task or to answer a set of questions are not appropriate at this stage, but may be introduced towards the end of the interview. The object of interviewing the patient is to assess disorders of language, memory, insight, and orientation. These quickly emerge in normal, social conversation.

After introducing himself (let us assume here a male interviewer and interviewee) and shaking hands the interviewer asks the patient's name, address, age, and date of birth. He proceeds to enquire about his spouse, children, children-in-law and grandchildren, his work and his health. He talks about photographs, cards, or other personal objects at the bedside. A question may be ventured on whether the patient knows why he is here and why he is being questioned; and he can be asked whether he thinks his memory is all right for his age.

This conversation usually allows the current mental state to be adequately assessed in a few minutes. It can be expanded if necessary to establish details of orientation, comprehension or recollection, noting the mechanisms which patients may adopt.

Evasion

Evasiveness may take the following forms.

- The patient appeals to a bystander to answer the question.
- The question is avoided by a smoke-screen of words, such as, 'Oh I know that all right', or 'I know perfectly well but it just won't come to me', or 'I could tell you if I wanted to'.
- The patient counter-attacks by saying, 'Why are you asking me these silly questions?' or 'You are trying to make out that I am crazy'.
- The patient side-tracks the question by taking over the conversation and

steering it into directions where he is more comfortable, e.g. by indulging in reminiscences.

- The patient 'confabulates' by deliberately or subconsciously fabricating what appear to be reasonable answers but which are drawn from another area of experience.

Language

Defects of language become apparent when the patient is given the opportunity of formulating sentences in response to open-ended questions. Words are misued, sentences are left incomplete, or speech becomes fragmented. This is to be differentiated from the dysphasia which occurs in stroke.

Memory

Defects of memory are detected and can be graded in severity by the answers to the questions about family. This information has been acquired and stored at different stages of life.

The name and date of birth have been known since early childhood, but age has to be updated each year. Gross inconsistency between date of birth and stated age indicates impaired brain function.

Patients who have remained in the same house for many years may record the correct address despite having brain failure, as the information was acquired when the brain was functioning normally. Patients who have moved house in recent years may give a former address. Knowledge of when they moved throws light on the possible time of onset of the brain failure.

The wife's first name has been used daily for many years. Her maiden name may have been little used since marriage. Failure or delay in recalling the maiden name suggests impaired brain function. Information about the difference in age between self and spouse has been known since before marriage. Gross inconsistency may be noted between the stated ages of each and the stated difference in ages.

The names of children have been known for half a century. Their married names have been known for a quarter of a century. The married name of a daughter and the maiden name of a daughter-in-law are impressed less firmly on the memory. Patients who can correctly and quickly state these names are unlikely to have severe brain failure, unless the circumstances of their lives threw them closely together, e.g. if the patient was living with a married daughter he might be expected to know her surname even if he had brain failure.

The names of grandchildren may have been known for a substantial time but their usage may have been limited if the patient rarely sees them. The patient who keeps an accurate check on the ages of children and grandchildren is unlikely to be suffering from severe brain failure.

Patients who have no spouse or children may be asked the names of their parents. The ability of a patient to recall and describe his occupation gives a fair idea of his memory, former social and intellectual level, and current skill in self-expression.

Insight

The questions on health, location, and memory aid understanding of how the patient sees himself. He may be unaware of being in hospital and deny the need, and he may express resentment, detachment, amusement, bewilderment, or wrath. He may deny memory impairment or admit that 'his memory isn't as good as it used to be', but dismiss this as a normal ageing phenomenon.

Orientation

Defects of orientation may emerge from responses to direct questions, or may be inferred from the patient's spontaneous remarks and behaviour. The commonly used direct questions are:

- What day is it, what month, what year, what time?
- Where are you?
- Do you know who I am?

These questions may however be resented or evaded, and it is better to say:

- Tell me what happens in this place.
- Tell me about the other people here.
- What have you done today?
- What do you usually do on a day like this?

Mental tests

The expression 'formal mental test' is used to describe a standard test procedure which gives a numerical score. These tests are quickly and easily performed. The following points are in their favour:

- The scores bear some relationship to the severity of the disease.
- Different tests give similar scores.
- Scores are fairly consistent on retesting.
- Clinical change is reflected in change in the score.

However, tests are reductionist in their attempt to express the width, depth, and subtlety of brain failure in a single number. They give little information about the quality of the patient's difficulty; and some patients find them irrelevant, threatening, or offensive. Tests may give misleadingly low scores in deaf, dysphasic, depressed, physically ill, and poorly educated patients, in those of low intelligence, and in non-native English speakers.

Three tests will be described.

Mental status questionnaire

The patient is asked 10 questions and gains one point for each correct answer. In one version the questions are:

- What is your name?
- What is your age?
- What is your address?
- What is the date of your birthday?
- What day is it?
- What month?
- What year?
- What time is it? (to the nearest hour)
- What is this place?
- What is my job?

A score of 7 or less is consistent with impaired brain function.

Set test

The patient is asked to name as many objects as he can, with a maximum of 10, in each of four consecutive categories or 'sets'. He is given as much time as is required to do so, but without further prompting. The maximum score is 40. The 'sets' are:

- colours;
- animals;
- fruits;
- towns.

A score of less than 25 in a person living at home or less than 15 in a hospital patient is consistent with impaired brain function.

Paired Associates

The patient is taught by two presentations to respond to a given word with an associated word, e.g. 'When I say "Knife", you say "Fork".' The stimulus words, in sets of three, are then offered three times each in random order, and the number of correct response words is counted. In the first part of the test the response word makes up an expected pair, in the second it makes an unexpected pair, and in the third it is unconnected in meaning. One mark is awarded for each correct response, with a maximum score of 27. The pairs are:

Knife Fork
East West
Hand Foot

Cup Plate
Cat Milk
Gold Lead

Grass Thimble
Chain Shoe
Stamp Rifle

In clinical practice with elderly, ill people only the first two parts of the test are necessary. The maximum score of 18 is attained easily by the mentally normal. The severely abnormal never grasp the association and score less than 9. Scores in between give a measure of the ability to remember the learned association and to disregard the expected one.

Mental tests are useful in patients with suspected focal disease and in epidemiological studies; but in everyday clinical work they add little to what is learned by listening to and observing the patient in his or her daily life.

12 *Intellectual impairment 3: examination, course, and management*

Physical examination and investigation

These are directed towards the detection of disease, which may be causal, contributory, consequential, or coincidental.

Causal

Alzheimer's disease and normal-pressure hydrocephalus have no characteristic physical signs. Vascular brain failure is often accompanied by focal neurological signs indicative of previous infarction. Focal neurological signs are to be sought also in primary and secondary brain tumours when these masquerade as brain failure. The characteristic physical signs of Parkinson's and Huntington's diseases are easily detected. Myoclonus may be found in any case of advanced brain failure, and is not diagnostic of Jakob–Creutzfeld disease.

Special tests are of value mainly in younger patients, and when there is some possibility of surgical intervention; but are not usually required in elderly patients with advanced brain failure. Radiography of the skull is of little diagnostic value. A computerized brain scan (CT) helps to exclude infarction, tumour, and normal-pressure hydrocephalus; but there are no characteristic findings in Alzheimer's disease. The characteristic EEG of Jakob–Creutzfeld disease may assist in the diagnosis of this rare condition.

Contributory

A vigorous quest should be made for accompanying infection, hypoxia, circulatory insufficiency, metabolic and endocrine disorders, malignant disease, anaemia, depression, and drug treatment. These all may contribute to the manifestations of brain failure; and treatment of them may improve brain function even in the presence of intrinsic brain disease.

Consequential

Malnutrition, anaemia, dehydration, constipation, incontinence, skin lesions, onychogryphosis, and burns are among the correctable consequences of lack of self-care.

Coincidental

Coincidental disease is liable to be overlooked in patients with brain failure, either because patients do not complain of the symptoms; or because the symptoms take the form of behaviour disturbance, as may occur in bladder and bowel disturbances, and in painful and infective conditions.

Sometimes coincidental and consequential conditions are looked upon as causal. This probably applies to some cases of megaloblastic anaemia and of thyroid and parathyroid disease, when identification and treatment of the condition has little or no beneficial effect on the progress of the brain failure.

Observing the patient

Fluctuation of mental state and variability of behaviour are characteristic of patients with brain failure. Patients often 'put on their best behaviour' for the benefit of the doctor or nurse, who goes away convinced that the family are exaggerating. Moments later or earlier a very different picture may have been evident.

Patients tend to be less able to understand the environment in the evening or at night, because of fatigue, low levels of illumination, or diurnal rhythms, and their behaviour at these times is disturbed. They may react in idiosyncratic ways with people who remind them of family members, school teachers, employers, army sergeants, or others from their past life.

The unfamiliar environment and routine of a hospital bewilders those whose powers of registration and discrimination are weakened.

Reports should be sought from staff or relatives on:

- dressing;
- feeding;
- washing;
- grooming;
- micturition and defaecation;
- movement through the ward area;
- shouting;
- offensive language;
- abuse;
- striking out;
- sleeping;

- behaviour to staff;
- behaviour to other patients;
- behaviour to visitors.

The course of brain failure

In extrinsic brain failure with a reversible cause, such as pneumonia, myocardial infarction, and toxic medication, brain function returns to normal with reversion of the cause. If brain function was previously impaired, it may remain more impaired than before the onset of the extrinsic cause. When the cause is not reversible, for example in hepatic and renal failure, and after hypoglycaemic and hypoxic brain damage, brain function does not revert to its previous level and may progressively decline.

In intrinsic brain failure with a reversible cause, e.g. after recent subdural haematoma and in meningitis, brain function may revert to normal; but it may remain impaired or decline after successful operation on longstanding subdural haematoma and normal-pressure hydrocephalus.

When the cause of brain failure is irreversible the course is prolonged and progressive. Alzheimer's disease is said to show steady, insidious, downward progression; multi-infarct disease is characterized by stepwise decline, each episode of infarction being related to a further aggravation of brain failure. But these stereotypes are deceptive, because the diseases may coexist; and because the course of each may be affected by contributory, consequential, and coincidental disease. Since there is at present no specific treatment for either disease, the main value in differentiating between them is the anxiety of relatives, who know of the familial incidence of Alzheimer's disease and fear that they may become affected themselves.

Stages

Intrinsic brain failure has a long course, which may be thought of as having four stages. These are called amnesic, behavioural, cortical, and decerebrate; and their respective durations are roughly four years, three years, two years and one year. The stages overlap and the durations are inexact, but the scheme gives an approximate idea of what to expect.

In the amnesic stage the major feature is impairment of recent memory. In the behavioural stage the picture is dominated by the patient's increasingly unpredictable and irresponsible behaviour and the family's attempts to contain it. In the cortical stage the spread of the pathology from the hippocampus and temporal lobe into the frontal, parietal, and occipital lobes of the brain leads to aparaxia, agnosia and aphasia, and perhaps incontinence. In the final decerebrate stage the patient becomes immobile, uncommunicative, doubly incontinent, and unable to recognize or to respond to others.

Management of brain failure

In the management of brain failure a carer with a healthy brain looks after a patient with an unhealthy brain. The patient cannot understand the behaviour of the carer; but the carer can and should understand the behaviour of the patient.

The patient's behaviour is determined always by his or her disordered perception of reality. Behaviour which might appear to be wayward, irresponsible, wicked, disgusting, thoughtless, or violent stems from the patient's misinformation, misperception, and misunderstanding.

Those who live with the patient through the early stages of the disease may expect logical behaviour and are tempted to correct illogical behaviour. For example, a patient who repeatedly asked the same question was told, with mounting impatience, 'Mother, I have already told you.' But the patient repeated the question, not in order to exasperate the daughter, but because she did not remember having asked it before. She was surprised and hurt by the daughter's unprovoked anger, and she in turn angrily denied having asked the question previously.

Experienced carers learn to expect 'illogical' behaviour, which stems logically from the patient's assumptions. They then adjust their responses to the patient's perceptions. In the example given, once the daughter accepts that the patient believes that each time she asks the question is the first time, she answers it as if it was the first time.

Some carers feel strongly that it is preferable to teach the patient to respond correctly, as otherwise incorrect behaviour is reinforced. This approach, which requires patience and tact, may be right for some people in the early stages of the disease. The patient is asked, 'I wonder if you remember asking that question before, and what answer I gave you?' If the patient pauses and reflects and admits the possibility, then this method should be pursued; but otherwise persisting in the belief that the patient can and should learn to behave as if he or she was not suffering from a major organic disease of the brain leads to frustration of both patient and carer.

The experienced carer aims to be objective, emotionally detached from the threatening aspects of the patient's behaviour, and able to recognize it as a logical response to the perceived situation. This is never easy and sometimes impossible. The carer's approach is to ask:

- What situation do I perceive?
- What situation does the patient perceive?

- Can I make him see the situation as I perceive it?
- If not, how can I best make his behaviour acceptable?

The following examples suggest alternative methods of handling situations.

Example 1

PATIENT: Where are you going?
DAUGHTER: To the hairdresser, Mother.

P: Where are you going?
D: To the hairdresser.

P: Where are you going?
D: To the hairdresser, I told you already.

P: No you didn't.
D: Yes I did, twice. You must have forgotten.

P: Where are you going?
D: Mother, I have told you three times.

Then the daughter stops and thinks. Where is this leading me? Mother obviously does not remember, but she remains anxious because she sees me going out and does not want to be left alone. Why am I arguing with her? Why don't I invite her to come with me?

Example 2

DAUGHTER: What are you doing?
PATIENT: Waiting for the bus.

D: But that isn't a bus-stop, it is a tree in the garden.
P: I'm waiting for the bus.

D: Where do you want to go to in a bus?
P: Home to Parliament Street.

D: But that is where you lived when you were first married. It has all been knocked down years ago. Come inside with me.
P: I am waiting for the bus to take me home.

D: But this is your home now. Don't you remember, you live with me now, your daughter, Joan.
P: No, I live in Parliament Street.

D: Come inside with me, you are getting all wet out here, I will make you a nice cup of tea.
P: Don't treat me as if I was a baby.

Then the daughter stops and thinks. The patient has made a perceptual error, seeing the similarity between a tree and a bus-stop, and overlooking their dissimilarity. She has made a conceptual error, incorporating the presumed bus-stop into a personal world that is centred on the powerful motive of Going Home. She has receded in time to a period when she was mistress of her home and family in Parliament Street. She does not recognize this argumentative, middle-aged woman as her daughter, and finds her inquisitive, interfering, and patronizing. Her location in time and space is not to be changed for the time being. Perhaps, thinks the daughter, it might be better for me with my normal brain to enter into her world, than to expect her with her abnormal one to enter into mine. So she tries again.

D: Hello, what are you doing?
P: I'm waiting for the bus.

D: So am I, do you know when it is due?
P: I have been waiting for a long time.

D: I think there must be a strike on today. Let's phone for a taxi. I'll tell you what, we'll go into the house and phone, and while we are waiting we will have a nice cup of tea together.
P: That is very kind of you.

Example 3

PATIENT: Hello Jeannie, how are you today? You are looking fine.
DAUGHTER: Mother, why do you call that lady Jeannie?

P: Because that is her name, Jeannie Atkins.
D: Mother, I keep telling you, that lady is a Mrs Merryweather, and you don't know her.

P: What do you mean I don't know her, I've known her all my life, haven't I Jeannie?
D: How could you have known her all your life when she only came in here last week and you had never seen her before?

P: We went to school together and we used to play in the street, and we even went to our first dance together in the old Palais. You remember that, don't you Jeannie?

Then the daughter stops and thinks. Why does the patient see the stranger as Jeannie?

D: Mother, tell me more about Jeannie Atkins.
M: She was my best friend, we went everywhere together.

D: What did she look like?
M: She was a beautiful girl.

D: What colour was her hair?
D: She had lovely red hair.
D: Mrs Merryweather has red hair.
M: Who is Mrs Merryweather?
D: That lady that you call Jeannie. Do you think you might be mixing her up with Jeannie Atkins because they both have red hair?
M: Well maybe.

Reality orientation: whose reality?

Relatives often ask whether they should attempt to correct faulty concepts, or 'go along' with them? The answer depends on what they hope to achieve, normality or tranquillity. Reality orientation teaches that patients with an abnormal concept of reality should have their misconceptions corrected and not reinforced. It recommends enrichment of the patient's environment by money, possessions, photographs, personal space, and newspapers; and it gives to the patient responsibility for activities which contribute to the household.

Reality orientation aims to keep patients functioning in a real if diminished world. It tries to find fragments of information which still have meaning for patients and which may draw them back to a heightened sense of reality. The third incident above is an example of reality orientation.

Patients who retain insight and a capacity to learn may benefit from correction of mistaken beliefs, provided that they are not flatly contradicted but are guided towards 'discovery learning'. There is, however, a danger of employing reality orientation inappropriately. The traditional 'Reality Orientation Board' to be found on the walls of hospital wards and residential homes, which states the day and the weather, is not found in real homes and merely reinforces the unreality of the environment.

Patients who lack insight and who have lost the capacity to learn either flatly reject any idea which is in conflict with their own beliefs or accept it for a fleeting moment then revert to their former misconception. Many resent being 'lectured', and indignantly deny that they have made a mistake or that they have already been told the answer to a question.

Whose reality?

The central assumption of reality orientation is that the patient should orient to the carer's reality. It seems more reasonable to require the carer to orient to the patient's reality.

With a little patience and imagination the carer learns to 'find' where the patient is located in time and place and to understand the patient's preoccupations. Once the carer has reached out and 'found' the patient, it is possible to conduct the patient back to the carer's reality, or to leave the patient where he or she was, but to understand and respond appropriately to his or her preoccupations.

Stress on carers

Patients with intrinsic brain failure spend most of the course of the disease at home, under the care of relatives, with support from the community services. This engenders severe stress, perhaps more severe than in any other illness because, as wives poignantly say, 'That's not my husband'. The unpredictable nature of the patient's behaviour, the coarsening of personality, the loss of companionship and reliability are bitter blows; and they are coupled with the constant anxiety that some dreadful but preventable accident will befall the patient.

The doctor's responsibilities include:

- to ensure that the diagnosis is right;
- to treat accompanying disease;
- to use drugs with circumspection, because of the sensitivity to their effects of patients with brain failure, and because of the patient's liability to error;
- to recall that patients may think that drugs are being given in order to poison them;
- to ensure that relatives receive allowances, domiciliary support, relief admission, day-hospital attendance, home appliances, psychogeriatric nursing, group discussions, counselling, literature, and any other help that is available;
- to ensure that relatives have a telephone number of a doctor, nurse, social worker, or volunteer on whom they can call when, or before, they are distraught;
- to see the patient and carer regularly, even if there is apparently not a lot to be done;
- to recall that some carers may not be far away from committing suicide or homicide.

Four don'ts for carers

Don't restrict
Don't conflict
Don't convict
Don't evict

Don't restrict

- A locked door is rattled.
- A table in front of a chair is beaten.
- Cot sides are climbed over.
- Sedative drugs are spat out.

Patients resent their 'gaolers'; escape from their prisons; fight against their drugs. They should be allowed to move in a safe environment; or they should sit in deck chairs or bean-bag chairs in rooms whose doors have double handles. The worst that can happen to wandering patients may be less bad than the consequences of preventing them wandering.

Don't conflict

The patient is just as convinced of being right as is the carer. To avoid conflict avoid correction.

Don't convict

When brain failure is perceived as a crime rather than as a disease the patient is in danger of being punished rather than treated.

Don't evict

When the behaviour of a patient becomes unacceptable a move is proposed to a different environment. In solving one problem this creates others. Will he be less or more troublesome in the new environment? Will he ever feel secure if he is not at home? Will the carer suffer feelings of guilt?

Ethical questions

The management of the patient with brain failure is as full of ethical problems as a pomegranate is of seeds. Two of the many issues will be raised here: autonomy, and giving and withholding treatment.

Autonomy

The civil and criminal law impose agreed restraints on the liberty of the individual in order to ensure the liberty of the community. But within the law there is free choice of whether, for example, to accept or refuse medical treatment, to enter or leave hospital, to agree or disagree to a recommendation for major or life-saving surgery. These rights are enjoyed by patients with brain failure, unless and until a doctor certifies that they should be deprived of them.

The law makes no special provision for people who are normal one minute and abnormal the next; for people who believe themselves to be fit and capable of self-care; for people who believe that their homes are ready and that their friends and relatives await them cheerfully; but who are, in fact, liable to inflict damage on themselves, their carers, and their environment.

The doctor may have to decide whether to respect the patient's autonomy and allow the patient to return home, which may endanger others; or to respect the autonomy of others and restrict the patient's liberty.

Giving or withholding treatment

The doctor accepts the responsibility always to do his or her best for the patient. This was interpreted as meaning preserving life at all costs. Does this principle extend to the unborn fetus, to the frozen embryo, to the congenitally deformed? Does it extend to the use of all possible forms of treatment or only to those normally available? Does it apply to prisoners who attempt to starve themselves to death?

The management of incidental disease among elderly patients with brain failure is a less public issue, but it raises the same questions, and it does so daily and universally. Should pneumonia in a terminally ill, aged patient with Alzheimer's disease be treated in the same way as in a mentally well, younger patient? And if so, what about myocardial infarction, or gangrene of the leg? Should patients who refuse food or who have difficulty in swallowing be fed by tube or intravenously? How often should tubes which have been pulled out by the patient be reinserted?

There is no answer to these questions; but staff and relatives are advised to discuss and state the reasons for their views before major decisions are taken, since what seemed at the time to be the right decision may not have the hoped-for outcome. If a decision is taken to amputate a limb and the patient dies; or not to amputate and the patient survives; if the patient is sent home and causes a fire; or is kept in hospital and escapes and is run over ... was the decision right at the time in spite of the outcome? The worst course is to take a decision which did not involve those who have to live with the consequences.

Notes and references

Classification

Standard diagnostic criteria for 'dementia' and 'delirium' are described in the third edition of the *Diagnostic and statistical manual of mental disorders*, known as DSM-III (American Psychiatric Association 1980). DSM-III has gained wide acceptance amongst psychiatrists, and has acquired its own 'expert system' for desk-top computer diagnosis (Plugge *et al.* 1990); but the criteria are not always easy to apply in clinical practice (Rabins 1989; Johnson *et al.* 1990).

Hachinski *et al.* (1974) distinguished between vascular and non-vascular types of dementia, while admitting the existence of overlap (Scheinberg 1988). The Hachinski scale, which awards points for signs thought to be characteristic of 'multi-infarct dementia', is least helpful when the distinction is clinically uncertain (Blessed 1980). There is a tendency for clinicians to overdiagnose Alzheimer's (Homer *et al.* 1988).

Wilcock (1988) found insufficient evidence to support the hypothesis that two distinct types of Alzheimer's disease can be distinguished, based on differences in age of onset, speed of progression, and clinical signs.

'Benign senescent forgetfulness'

'Normal' memory impairment in later life is to be distinguished from progressive dementia (West 1985). The 'normal' person may be intensely worried by relatively mild impairment (Yesavage *et al.* 1982).

Rating scales

Ever since Blessed *et al.* (1968) showed that there was an association between mental test scores in late life and pathological changes in the brain after death, a large number of functional assessments and tests have become available (Wattis and Hindmarch 1988). These include very brief questionnaires, with no more than 10 simple questions on current information (Kahn *et al.* 1960); the use of educational children's toys (Isaacs and Walkey 1964); objective measures of brain function using auditory-evoked potentials and automated and computerized, self-administered tasks (Wright *et al.* 1988; Alexander and Davidoff 1990); and comprehensive batteries covering many aspects of behaviour and function, such as the Stockton rating scale (Meer and Baker 1966; Gilleard and Pattie 1977); the Clifton Assessment Procedure for the Elderly (Pattie 1981); the Mini mental state (Folstein *et al.* 1965); the Geriatric mental state (Copeland 1976), and the Older American Resources and Services (George and Fillenbaum 1985). Israel *et al.* (1984) list 245 published 'instruments' for geriatric assessment. Greene *et al.* (1982) devised scales for measuring strain on relatives. In the United Kingdom most experience has been gained with a short empirical, bedside test (Hodkinson 1972), and with the Set test (Isaacs and Akhtar 1972); while the quick and informative Paired Associate test (Isaacs and Walkey 1964) has not been widely adopted.

Most tests have been shown to correlate well with one another and with DSM-III criteria (Kafonek *et al.* 1989); but omit much subtle information which can be obtained from observation of the patient's dress, language, and behaviour (Jones and Williams 1988).

Symptomatology

The manifestations of dementia include disturbance of language function (Code and Lodge 1987; Powell and Cummings 1988); visual impairment (Steffes and Thralow 1987); hearing impairment (Peters *et al.* 1988); impaired ability to process heard material (Grimes *et al.* 1985); sleep disturbance (Lowenstein 1982; Prinz *et al.* 1982, 1990; Vitiello *et al.* 1990); weight loss (Singh *et al.* 1988); depression, apathy, and ritual behaviour (Teri *et al.* 1989); violent behaviour (Petrie *et al.* 1982; Mentes and Ferrario 1989); manual removal and smearing of faeces (Begg and McDonald 1989). When dementia is accompanied by hearing loss the rate of cognitive decline increases (Peters *et al.* 1988).

Families designated 22 different manifestations of the disease as the 'biggest problem' which they faced (Rabins *et al.* 1982).

Differential diagnosis

Brain tumours are excluded by the presence of fits, focal signs, and the rapidity of progression (Godfrey and Caird 1984). About one patient in 10 who presents with dementia has the other two symptoms of the 'triad' of normal-pressure hydrocephalus, namely urinary incontinence and gait disturbance; and about one in 10 of these has the CT signs (Mulrow *et al.* 1987). Bypass operation is not recommended if the dementia is severe, as clinical improvement is unlikely.

Binswanger's disease, or subcortical arteriosclerotic encephalopathy, is diagnosed by nuclear magnetic resonance imaging; but is not clearly distinguishable clinically from other forms of dementia.

The question of how extensively elderly people with dementia need to be investigated was discussed at a Consensus Conference (1987). Paramount emphasis is placed on the history. The belief that a substantial proportion of dementias are 'reversible' (Smith and Kiloh 1981) owes much to short-term reports from specialist units dealing with a relatively young population (Clarfield 1988). Even when the dementia is caused or aggravated by drugs, depression, or metabolic disorder, sustained improvement is rare (Larson *et al.* 1984).

Management

The use of reality orientation (RO) in demented patients led to small, positive changes in behaviour (Woods and Holden 1982); but it is uncertain that this is due to the RO programme (Greene *et al.* 1983). The book *The 36 hour day* (Mace and Rabins 1981) gives excellent guidance to relatives.

Late paraphrenia

This condition is a variant of schizophrenia, commencing after the age of 60 (Roth 1955; Post 1966). It tends to occur in people of cold and distant personality, many of whom have never married. Its manifestations include fantastic, grandiose, or persecutory delusions, delusions of reference, and hallucinations in a setting of clear consciousness, and in the absence of dementia, cerebrovascular disease, alcoholism or affective disorder (Hymas *et al.* 1989).

References

Alexander, J. E. and Davidoff, D. A. (1990). Psychological tests, computers and aging. *International Journal of Technology and Aging,* **3,** 47–56.

American Psychiatric Association (1980). *Diagnostic and statistical manual of mental disorders,* (3rd edn) APA, Washington DC.

Begg, A. H. and McDonald, C. (1989). Scatolia in elderly people with dementia. *International Journal of Geriatric Psychiatry,* **4,** 53–4.

Blessed, D. G., Tomlinson, B. E., and Roth, M. (1968). The association between quantitative measures of dementia and of senile changes in the cerebral grey matter of old people. *British Journal of Psychiatry,* **114,** 797–811.

Clarfield, A. M. (1988). The reversible dementias: do they reverse? *Annals of Internal Medicine,* **109,** 476–86.

Code, C. and Lodge, B. (1987). Language in dementia of recent referral. *Age and Ageing,* **16,** 366–72.

Consensus Conference (1987). Differential diagnosis of dementing diseases. *Journal of the American Medical Association,* **258,** 3411–16.

Copeland, J. R. M. (1976). A semi-structured clinical interview for the assessment of mental state in the elderly: the geriatric mental state schedule. I. Development and reliability. *Psychological medicine,* **6,** 439–49.

Folstein, M. F., Folstein, S. E., and McHugh, P. R. (1975). Mini mental state: a practical method for grading the cognitive state of patients for the clinician. *Journal of Psychiatric Research,* **12,** 189–98.

Gilleard, C. J. and Pattie, A. H. (1977). The Stockton geriatric rating scale: a shortened version with British normative data. *British Journal of Psychiatry,* **131,** 90–4.

Godfrey, J. B. and Caird, F. I. (1986). Intracranial tumour in the elderly: diagnosis and treatment. *Age and Ageing,* **13,** 152–8.

Greene, J. G., Smith, R., Gardiner, M., and Timbury, G. L. (1982). Measuring behavioural disturbance of elderly demented patients and its effects on relatives: a factor analytic study. *Age and Ageing,* **11,** 121–6.

Greene, J. G., Timbury, G. L., Smith, R., and Gardiner, M. (1983). Reality orientation with elderly patients in the community: an experimental evaluation. *Age and Ageing,* **12,** 38–43.

Grimes, A. M., Grady, C. L., Foster, N. L., Sunderland, T., and Paronas, N. J. (1985). Central auditory function in Alzheimer's disease. *Neurology,* **35,** 352–8.

Hachinski, V. C., Lassen, N. A., and Marshall, J. (1976). Multi-infarct dementia: a cause of mental deterioration in the elderly. *Lancet,* **ii,** 207–10.

Hodkinson, H. M. (1972). Evaluation of a mental test score for the assessment of mental impairment in the elderly. *Age and Ageing,* **1,** 233–8.

Homer, A. C., Honavar, M., Lantos, P. L., Hastie, I. R., Kellett, J. M., Millard, P. H. (1988). Diagnosing dementia: do we get it right? *British Medical Journal,* **297,** 894–6.

Hymas, N., Naguib, M., and Levy, R. (1989). Late paraphrenia: a follow-up study. *International Journal of Geriatric Psychiatry,* **4,** 23–9.

Isaacs, B. and Akhtar, A. J. (1972). The set test; a rapid test of mental function in old people. *Age and Ageing,* **1,** 222–6.

Isaacs, B. and Walkey, F. A. (1964). Measurement of mental impairment in geriatric practice. *Gerontologia Clinica,* **6,** 114–23.

Israel, L., Kozarevic, D., and Sartorius, N. (1984). *Source book of geriatric assessment.* Karger, Basel.

Johnson, J. C. *et al.* (1990). Using DSM-III criteria to diagnose delirium in elderly general medical patients. *Journal of Gerontology,* **45,** M113–19.

Jones, T. V. and Williams, M. E. (1988). Rethinking the approach to evaluating mental functions of older persons: the value of careful observation. *Journal of the American Geriatrics Society,* **36,** 1128–34.

Kafonek, S., Ettinger, W. H., Roca, R., Kittner, S., Taylor, N., and German, D. S. (1989). Instruments for screening for depression and dementia in a long term care facility. *Journal of the American Geriatrics Society,* **37,** 29–34.

Kahn, R. L., Goldfarb, A. I., Pollock, M., and Peck, A. (1960). Brief objective measures for the determination of mental status in the aged. *American Journal of Psychiatry,* **117,** 326–8.

Lowenstein, R. (1982). Disturbances of sleep and cognitive functioning in patients with dementia. *Journal of the Neurobiology of Ageing,* **3,** 371–7.

Mace, N. and Rabins, P. (1981). *The 36 hour day: a family guide to coping for persons with Alzheimer's disease, related dementing illnesses and memory loss in later life.* Johns Hopkins Press, Baltimore.

Meer, B. and Baker, J. A. (1966). The Stockton geriatric rating scale. *Journal of Gerontology,* **21,** 392–403.

Mentes, J. C. and Ferrario, J. (1989). Calming aggressive reactions: a preventive program. *Journal of Gerontological Nursing,* **15,** 22–7.

Mulrow, C. D., Feussner, J. R., Williams, B. C., and Vokaty, K. A. (1987). The value of clinical findings in the detection of normal pressure hydrocephalus. *Journal of Gerontology,* **42,** 277–9.

Pattie, A. H. (1981). A survey version of the Clifton Assessment Procedure for the Elderly. *British Journal of Clinical Psychology,* **20,** 173–8.

Peters, C. A., Potter, J. F., and Scholer, S. G. (1988). Hearing impairment as a predictor of cognitive decline in dementia. *Journal of the American Geriatrics Society,* **36,** 981–6.

Petrie, W. M., Lawson, E., and Hollender, M. (1982). Violence in geriatric patients. *Journal of the American Medical Association,* **284,** 443–4.

Plugge, A., Verhey, F. R. J., and Jolles, J. (1990). A desktop expert system for the differential diagnosis of dementia. *International Journal for Technology Assessment in Health Care,* **6,** 147–56.

Post, F. (1966). *Persistent persecutory states of the elderly.* Pergamon, Oxford.

Powell, A. L. and Cummings, J. L. (1988). Speech and language alterations in multi-infarct dementia. *Neurology,* **38,** 717–19.

Prinz, P. N. *et al.* (1982). Changes in the sleep and waking EEG's of undemented and demented elderly subjects. *Journal of the American Geriatrics Society,* **30,** 86–93.

Prinz, D. N., Vitiello, M. V., Raskind, M. A., and Thorpy, M. J. (1990). Geriatrics: sleep disorders and aging. *New England Journal of Medicine,* **323,** 520–6.

Rabins, P. V. (1989). Published American definitions of dementia in Alzheimer's disease: a mid-1980's view. In *Innovative trends in psychogeriatrics* (ed. J. Wertheimer, P. Baumann, M. Gaillard, and P. Schwed), pp. 70–3.

Rabins, O. V., Mace, N. L., and Lucas, M. J. (1982). The impact of dementia in the family. *Journal of the American Medical Association,* **248,** 333–5.

Roth, M. (1955). The natural history of mental disorder in old age. *Journal of the Mental Sciences,* **101,** 281–301.

Scheinberg, P. (1988). Dementia due to vascular disease: a multifactorial disorder. *Stroke,* **19,** 1291–9.

Singh, S., Mulley, G., and Losowsky, M. S. (1988). Why are Alzheimer's patients thin? *Age and Ageing,* **17,** 21–8.

Smith, J. S. and Kiloh, L. J. (1981). The investigation of dementia: results in 200 consecutive admissions. *Lancet,* **ii,** 824–7.

Steffes, R. and Thralow, J. (1987). Visual field limitation in the patient with dementia. *Journal of the American Geriatrics Society,* **35,** 198–204.

Teri, L., Borson, S., Kiyakh, A., and Yamagishi, M. (1989). Behavioural disturbance, cognitive dysfunction and functional skill: prevalence and relationship in Alzheimer's disease. *Journal of the American Geriatrics Society,* **37,** 109–16.

Vitiello, M. V., Prinz, P. N., Williams, D. E., Frommlet, M. S., and Ries, R. K. (1990). Sleep disturbances in patients with mild stage Alzheimer's disease. *Journal of Gerontology,* **45,** M131–8.

Wattis, J. P. and Hindmarch, I. (1988). *Psychological assessment of the elderly.* Churchill Livingstone, Edinburgh.

West, R. (1985). *Memory fitness over 40.* Triad Publishing Co., Gainesville, FL.

Wilcock, G. K. (1988). Recent research into dementia. *Age and Ageing,* **17,** 73–86.

Woods, R. T. and Holden, U. P. (1982). Reality orientation. In *Recent advances in geriatric medicine,* Vol. 2 (ed. B. Isaacs), pp. 181–99.

Wright, G. M., Scott, L. C., Richardson, C. E., Rai, G. S., and Exton-Smith, A. N. (1988). Relationship between the P-300 auditory event-related potential and automated psychometric tests. *Gerontology,* **34,** 134–8.

Yesavage, J. A., Rose, T. L., and Spiegel, D. (1982). Relaxation training and memory impairment in elderly normals: correlation of anxiety ratings and recall impairment. *Experimental Aging Research,* **8,** 195–8.

Aphorisms

Of brain failure

- Failure of the heart is called heart failure; failure of the brain is called brain failure. 'Brain failure' is not a diagnosis, any more than 'heart failure' is a diagnosis; but it leads to a quest for a diagnosis.
- The heart is a pump, and when it fails it is a failed pump. The brain is a mystery, and when it fails it remains a mystery.
- If the patient gives a reliable history he is not suffering from brain failure. If he is suffering from brain failure he does not give a reliable history.
- The brain is the organ of thought; brain failure is failure of thought. The brain is the organ of perception; brain failure is failure of perception. The brain is the organ of behaviour; brain failure is failure of behaviour.
- Late brain failure is perceived as disease and is treated with compassion; early brain failure is perceived as aberration and is treated with passion.
- Early brain failure should not be described as 'mild' or 'moderate'; but as 'wild' or 'immoderate'.
- The evidence of early brain failure is:
  an increase in the telephone bill;
  an increase in the gas bill;
  an increase in the grocer's bill;
  an increase in the locksmith's bill.

Of the failing brain

- The patient with brain failure retains intelligence, logical thought, self-awareness, and the instinct for self-preservation.
- The normal brain perceives the dissimilarities of the similar; the failing brain perceives the similarities of the dissimilar.
- New percepts challenge old concepts. The normal brain modifies concepts; the failing brain modifies percepts.
- In the normal brain, preconceptions yield to evidence; in the failing brain, evidence yields to preconceptions.
- For the normal brain here is here and now is now. For the failing brain here is there and now is then.
- The normal brain operates logically from its here and now; the failing brain operates logically from its here and now.
- The normal brain can comprehend that the here of the failing brain

is not here and its now is not now. The failing brain cannot comprehend that the here of the normal brain is here and its now is now.

- The normal brain associates effect with cause; the failing brain dissociates effect from cause.
- The normal brain surveys time past, time present, and time future. The failing brain stands on a 'two-minute island of time'.
- The normal brain predicts future events from past experience. The failing brain does not predict future events from past experience.
- The normal brain perseveres; the failing brain perseverates. To persevere is to go on until you do it right; to perseverate is to go on doing it wrong.
- Orientation is congruence between concept and percept. When percept and concept are compatible, orientation is correct and stable. When percept and concept are incompatible, orientation is incorrect and unstable.

Of behaviour

- Patients with brain failure don't do what they can't do, and do wrong what they can do.
- Behaviour appears purposeless when its purpose is unknown.
- Wandering in brain failure is of four types:
  amnestic—searching but forgetting for what;
  anachronistic—engaged in past activity;
  anxiolytic—relieving anxiety;
  atavistic—back to the beginning.
- A reiterated question means an unregistered response. A reiterated behaviour means an unsatisfied need.
- Confabulation clothes impoverished thought in rich language, like an aged pauper in faded finery.
- The patient with brain failure exploits the ambiguities of language for the concealment of ignorance. So does the student at a viva and the Minister at the Dispatch Box.
- Patients with brain failure know no connection between me and mine or between you and yours. They give what is not theirs to give, and take what is not theirs to take. The first is called waste, the second theft.

Of mental tests

- Evaluating the brain by a score is like evaluating a painting by its price.

- Ignorance of the name of the Prime Minister is a poor indicator of brain failure; ignorance of the name of a grandchild is a good indicator of brain failure.
- A test is not objective when its content is subjectively selected; it is not consistent when the same questions are asked in different ways by different people; and it is not validated merely because it gives the same result as another, non-validated test.

Of management

- The Law of Brain Failure states that to every action there is an equal and opposite reaction; and to every motion there is an equal and opposite commotion.
- Caring for patients with brain failure means the investment of time, energy, and emotion, at low rates of interest and without capital appreciation.
- The object of good management is the avoidance of conflict. When patient and carer experience different realities there is conflict. When patient and carer experience the same reality there is no conflict.
- Reality orientation is the adjustment of the carer to the patient's reality, rather than the adjustment of the patient to the carer's reality.
- In hotels the staff adjust their hours to the needs of the guests. In hospitals the guests adjust their hours to the needs of the staff.
- For those who go to bed at 11 p.m., 3 a.m. is the middle of the night. For those who go to bed at 7 p.m., 3 a.m. is the start of the day.
- The carer of the patient with brain failure requires:
  to locate the patient in time and space;
  to share the patient's anxiety;
  to ignore her own emotions;
  to escort the patient out of bewilderment.
- In the management of brain failure observe the Four Don'ts:
  Don't restrict—the patient has his freedom.
  Don't contradict—the patient has his beliefs.
  Don't convict—errors are not crimes.
  Don't evict—the patient has his rights.

13 *Stroke 1: neurology*

Stroke illness in old age is so common and so disabling that it can be considered as one of the Giants of Geriatrics. However, only those aspects of stroke will be dealt with which require special emphasis in the management of elderly stroke patients in hospital.

This chapter presents a simplified account of the neurological and clinical features of stroke. History, examination, and rehabilitation are dealt with in the succeeding chapter; in Chapter 15, aphasia is treated as a Giant in its own right. The aetiology, pathology, epidemiology, prognosis, and medical treatment of stroke, and the subject of transient ischaemic attacks, are not dealt with.

The neurology of stroke

Simplified concepts, metaphors, and non-technical language are used in this presentation. Technical descriptions are available in standard neurological texts.

The brain may be likened to a factory which operates at three levels. These will be called operative, or 'shop floor'; administrative, or 'board room'; and communicative, or 'post room'. The 'operative' or 'shop floor' function means the ability to perform a defined motor or sensory task, such as moving a limb or feeling a pinprick. The 'administrative' or 'office' function refers to control of the planning and execution of a programme of movement, like putting on a skirt or pair of trousers; or analysing and giving meaning to a stream of percepts, like recognizing the material of which the garment is made. The communicative or 'post room' function is responsible for 'internal mail', i.e. transferring information and instructions from one part of the brain to another.

A stroke is like a bomb in the brain, causing damage at the site of the 'explosion' and disrupting communication with undamaged parts. Loss of neuronal tissue in stroke resembles bomb damage to the factory. When the shop floor is damaged, production in that part of the factory ceases until repairs can be made and new staff brought in. When the offices are damaged, production may continue for a while, but it is unplanned and unsupervised, and the products may be defective. When both shop floor and office are damaged, routine work is not done, and there is no mechanism for reorganizing the work, engaging new staff, or redeploying and retraining existing staff. When the post room is damaged, one part of the organization does not know what the other parts are doing.

In stroke illness, loss of operative function produces paralysis or loss of sensation, but the brain as a whole carries on until the damage is repaired or compensated for. When administrative functions are lost, movement and sensation may still be present, but they do not combine to allow functions to be performed. Loss of communication leads to 'disconnection' and autonomous behaviour by different parts of the brain, which may appear illogical or inconsistent.

Broadly speaking, loss of operative function is associated with lesions in the brain-stem and internal capsule; loss of administrative function with lesions in the cortex; and loss of communication with lesions of the white matter. But in brain damage, as in bomb damage, it is not easy to relate the disturbed function to the location of the damage.

The two hemispheres of the brain may be compared to two factories belonging to the same company. They manufacture the same product but distribute it to different regions. The shop floors of the two factories are identical, but their offices are different. Some administrative tasks are performed in both factories, some only in one, and some on a larger scale in one than in the other. In the same way the two cerebral hemispheres have areas of common function and areas of specialization, but neither is 'dominant'. For some administrative functions, such as the interpretation of tactile stimuli, damage to the specialized areas of one hemisphere is partly compensated by the other hemisphere. For other functions, notably language and the perception of space, the administrative function is almost wholly confined to one hemisphere; so that the left hemisphere can almost be thought of as the 'speech brain' and the right as the 'space brain'.

Clinical Manifestations

The terms used for the varied manifestations of stroke are reduced in Table 13.1 to those which refer to loss or impairment of operative and administrative function in relation to movement, sensation, and vision. A separate section is devoted to disconnection. These divisions are artificial, since all elements interact, but the terminology helps understanding.

Table 13.1 Manifestations of stroke

	Operative	Administrative
Movement	Hemiplegia, hemiparesis	Apraxia, dyspraxia
Sensation	Hemianaesthesia	Agnosia
Vision	Hemianopia	Visual agnosia

Movement

Damage at the 'operative' level of movements of the limbs and trunk leads to *hemiplegia* or loss of the ability to execute voluntary movement of one or both limbs on one side of the body. Impairment of this ability is called *hemiparesis*.

Damage at the 'administrative' level leads to *apraxia* or *dyspraxia*, in which there is loss or impairment, respectively, of the ability to organize and carry through correctly a complex sequence of movements, such as putting on a pair of trousers. The individual components of the complex can be separately performed, but they cannot be put together to achieve the purpose. In some cases the movement can be performed in imitation of another person, but cannot be initiated voluntarily or on verbal instruction.

The motor functions of swallowing, articulating consonants, and creating the vowel sounds are organized bilaterally in the brain-stem. Damage at the executive or administrative level leads to dysphagia, dysarthria, and dysphonia.

Sensation

Loss of touch, pain, and temperature perception at 'operative' level in the trunk and limbs on one side of the body is called 'hemianaesthesia' or 'hemianalgesia'. There is no separate term for partial loss. Unilateral loss or impairment of hearing, taste, and smell at 'operative' level result from lesions in the brain-stem.

Disturbed sensory function at administrative level is called *agnosia*, with no term to distinguish between loss and impairment. Agnosia occurs in relation to all forms of sensation. Agnosia for taste and smell are difficult to detect and are not prominent in stroke illness.

Tactile agnosia

In tactile agnosia the patient cannot interpret touch, and is thus liable to drop objects or to be unable to use them. Feeding and dressing are difficult because the patient cannot recognize the tactile patterns of cutlery, crockery, and clothing.

Proprioceptive agnosia

Agnosia for proprioceptive and kinaesthetic sensation means that the patient cannot comprehend the subconscious feelings which normally provide information about the state and position of the limbs and trunk and their movement in space. The patient loses control over his or her own movements. The leg swings wildly during walking, or stamps on the floor, or is left behind when the next pace is due. The arm lies 'neglected' at the side, or crushed between the body and the chair. Bizarre beliefs develop about the affected parts of the body and their relation to unaffected parts. Patients may deny the existence of a limb, or believe that they have a third upper limb. There is no short name for this condition.

Auditory agnosia

The patient hears but does not understand what is heard. Auditory agnosia differs from aphasia, which is specific to the understanding of language. The patient with auditory agnosia can comprehend speech but cannot recognize the significance of sounds such as a dripping tap or a whistling kettle.

Other forms of agnosia are described as separate clinical entities with long Greek names; but are treated here as combinations of executive and communicative disorders of the basic senses.

Vision

At the operative level, loss of the ability to see the right or left half of the visual field is called 'homonymous hemianopia'. Hemianopic patients compensate for their disability. They treat the half field which they see as if it was the whole visual scene, as a normal person does when he or she closes one eye.

Visual agnosia is impaired ability to interpret the perceived visual scene. In unilateral visual agnosia the visual world perceived by the unaffected hemisphere is normal, while that perceived by the affected hemisphere is absent or suppressed. The patient sees two half-fields, one of which has nothing in it. He ignores the existence of everything in the affected half-field, but reacts normally to everything in the unaffected one. He bumps into beds and chairs in the absent visual field, he eats the food in one half of the plate and leaves the rest untouched; he can tell the time if the hands of the clock are in the unaffected field but not if they are in the affected one; he reads half a line of print and then jumps to the next line. He responds to what he sees, and makes no attempt to 'fill in' what he does not see.

Bilateral visual agnosia is sometimes called 'occipital blindness'. In severe cases the patient may see nothing at all or only a featureless world. In incomplete cases, aspects of the visual world which attract the patient's interest or which are easily recognized are perceived, and others are not. This may create the erroneous impression that the patient 'can see when he wants to'.

Disconnection

Disconnection can result from a lesion involving the corpus callosum, which results in the right half of the brain not knowing what the left half is doing; or from lesser lesions of the white matter, with varied manifestations. The following case reports are examples of disconnection.

- A man of 69 with severe left hemiplegia described himself as having nothing wrong with him. He admitted that his left arm and leg were totally paralysed but had no doubt that he could drive a car. There was disconnection between his present and his previous image of himself.
- A man of 74 with left hemiplegia failed to learn to walk despite many weeks of treatment. When the physiotherapist instructed him what to do, he repeated the instructions precisely, then proceeded to do something quite different. There was disconnection between his words and his deeds, no doubt because of damage to the corpus callosum. When the patient was asked, 'Can you move your left arm' he said 'Yes', and lifted the paralysed arm with the normal one. There was disconnection between the transitive and intransitive meanings of the verb 'to move'.
- A mentally normal woman of 79 with left hemiplegia announced that she had two left arms, her own which was under the covers and which she could not see, and the one on the bed which she could see but could not feel. There was disconnection between her visual percept of a limb on top of the bed, and her proprioceptive memory of a limb under the covers. It sems that these two concepts could coexist, because they were disconnected from that part of the brain which matches current percepts against past knowledge and experience, and rejects false conclusions.

Other manifestations of stroke

A few of the varied manifestations of stroke will be described.

Muscle tone

Contrary to classical teaching, reduced muscle tone and loss of tendon reflexes on the affected side are common, especially in the early stages of stroke. Increased muscle tone and increased tendon reflexes may develop if the limbs are badly positioned.

Cerebellum and extrapyramidal system

A lesion involving the cerebellum causes balance disorder and incoordination. One involving the basal nuclei and the extrapyramidal system may cause mild or severe writhing and twisting movements on the affected side.

Monocular blindness

Temporary or permanent occlusion of the retinal artery causes loss of vision in one eye, but the visual field is unrestricted. This differs from homonymous hemianopia (see above), which the patient may erroneously describe as loss of sight in one eye.

Dysarthria and dysphonia

Speech is slurred when the cerebellum and its connections are affected, and explosive or whispered when the lesion is in the brain-stem.

Dysphagia

Dysphagia may result from a brain-stem lesion; or it may be present in the early stages of stroke illness, particularly when consciousness has been impaired.

Right–left dissociation

Some stroke patients are unable to distinguish right from left. When they are asked to move or to indicate one or other arm or leg, they choose the wrong one or do not know which to choose. Some normal people experience similar difficulty; but the dissociation found in stroke illness is of greater degree, and is usually accompanied by incomprehension of other spatial relationships and of the names of parts of the body.

Brain failure

Patients who have sustained multiple strokes can be described as having multi-infarct disease, which is one of the causes of intrinsic brain failure (see Chapter 10). The disturbances of thought and behaviour associated with a single stroke result from disconnection between functional areas of the brain; the term 'brain failure' is not appropriate.

Emotional lability

Some stroke patients have an uncontrollable tendency to burst into tears or laughter for no adequate reason, and to their great embarrassment. This is sometimes erroneously called 'emotional lability'. The term 'emotional dissociation' is more appropriate, since it is a manifestation of disconnection or dissociation between the expression of emotion and the experience of emotion. The patient weeps without being sad, or laughs without being glad.

Body image

Stroke patients who cannot recognize or name parts of their own bodies, or the parts of the body shown in drawings or in manikins, are sometimes said to suffer from disturbances of the body image. This condition can be thought of as due to loss of touch and proprioception, agnosia for touch and proprioception, dyspraxia, dysphasia, or any mixture of these.

Parietal lobe syndrome

This name is applied to patients who exhibit disturbances of the perception of space and spatial relationships; who put their clothes on in the wrong order; and who make mistakes in tests of visuospatial function, such as drawing arrows, stars, or human figures. These clinical features are not consistently defined; there is no close association between the symptoms and the location of the lesion; and without a scan it is not possible to know where the lesion is. The term should not be used.

Depression

Depression is common after stroke, which seems natural enough in view of the age of the patients, the frightening manifestations of the disease, and the great effort required to return to normal living. Antidepressant therapy is often helpful. The relationship between life-events, brain damage, drugs, mood, and the criteria for defining depression is so complex that some doctors are tempted to say that if the patient appears depressed and there are no contraindications to the use of antidepressant drugs, then s/he should be given the drugs, and if s/he gets better s/he must have been depressed. The matter is discussed in Chapter 18.

Neglect of a limb

This phrase is used in two senses, non-use and unawareness. Because of this ambiguity it is better to avoid the phrase.

Non-use of a limb which is not paralysed is a manifestation of apraxia. The limb may participate in a movement which the patient has not tried to produce, but it cannot be moved at will. This gives the impression that the patient is not trying, or that s/he is neglecting to use the limb; but the true facts are summarized by saying that:

> Apraxia is when you can't do what you can do when you try; but you can do what you can't do when you don't try.

The other sense of neglect of a limb is when the patient looks away from the affected limb, allows it to drop uselessly at the side, and seems unaware of its existence. This is due to loss of tactile and proprioceptive sensation at both operative and executive level. The word 'neglect' is also used in connection with the apparent neglect of objects in one visual field which characterizes unilateral visual agnosia.

14 *Stroke 2: history, examination, and rehabilitation*

Taking the History

Which side?

There are many difficult challenges in stroke management, but one easy one is to record unambiguously the side of the stroke. The side of a stroke is the side which is clinically abnormal. The fact that the lesion may be in the opposite side of the brain should not confuse the issue. Much of the trouble is caused by replacing the excellent clinical term 'stroke' with the ambiguous acronym CVA (cerebrovascular accident); since it is unclear whether accidents occur on the side of the cause or the effect. A patient with right hemiplegia has a right-sided stroke, and a patient with left hemiplegia has a left-sided stroke.

Other difficulties

The following more serious factors may cause difficulty in obtaining a reliable history from stroke patients.

1. Clouding or loss of consciousness at the onset, with amnesia for the events of that period.
2. Impaired ability to recall and organize information as a consequence of the current or previous strokes.
3. Language disturbances, either gross, as in dysphasia, or minor and subtle in non-dysphasic patients with left-hemisphere lesions.
4. Difficulty in describing the unfamiliar experiences which characterize loss of motor, sensory, and language skills.
5. Patients who have sustained a right-hemisphere lesion may be garrulous, joky, optimistic, and dismissive of the severity of their symptoms.

History of the current stroke

As in all conditions of sudden or rapid onset, the patient is asked to describe what happened, when it happened, where s/he was when it happened, what s/he was doing when it happened, and why in his or her opinion it happened. Intelligent, articulate patients spontaneously describe their altered function, sensations, and emotions. Those with lesser language skills may overlook significant symptoms, or may describe them in inappropriate language.

The following symptoms are enquired about, if they are not mentioned by the patient:

- weakness or paralysis of a limb or of one side of the face;
- distortion of facial appearance;
- looseness of dentures, because of facial weakness;
- 'pins and needles' and other abnormal sensations;
- seeing double;
- difficulty in swallowing saliva;
- difficulty in drinking;
- difficulty in articulation;
- impairment of balance;
- inability to find words;
- inability to comprehend what other people are saying;
- inability to read and to count;
- inability to write and draw;
- a sense of unfamiliarity;
- failure to recognize parts of the body.

Patients with motor disturbances may speak of the affected limb as being 'heavy' because it requires a surprising amount of effort to move it; or as 'naughty' because it refuses to obey their will; or as 'lazy' because it responds slowly.

Patients with sensory loss at 'operative' level (the levels are described in Chapter 13) describe the limb as 'numb' or 'dumb' or 'cold' because they do not know where it is or what it is doing or feeling. Some use expressions like water dripping or a wind blowing through the limb.

Those who experience loss of sensation and proprioception at 'administrative' level may think of the limb as no longer part of their body. They give it a name, like 'Lazy Louis' or 'Useless Eustace', speak to it as if it was a naughty child or a pet dog, and spank it because it does not do what it is told. They may say that the limb is made of wood; that it belongs to someone else; or that it was put there by the physiotherapist. Some believe that they have two left arms. Patients who experience these bizarre sensations may fear to speak of them in case they are dismissed as 'crazy'. Patients are therefore asked, 'Have you experienced any symptoms which you cannot understand, or which seem crazy to you?'

Patients may be unaware of having hemianopia or visual agnosia; or they may speak of blurring of vision or a greyness coming over their eyes.

Psychological history

Patients and relatives are interviewed together, so that information and emotions are shared. Many are frightened and bewildered by a stroke. Few know what the word really means, and some fear that the disease is fatal or that permanent invalidism is certain.

The opinions of relatives on the cause and cure of the stroke may differ from those of the patient. Both are asked what they think caused the stroke; what the stroke has done to them; and what are their hopes of recovery. They are asked to name anyone who has had a stroke and to describe the outcome.

Many recount a stressful event which preceded the stroke and which they believe must have caused it. The patient may accuse the spouse, or the spouse may accuse the patient, of being to blame for the stroke. ('If only he had listened to me this would never have happened.'). When the stroke occurs during or soon after sexual intercourse it is very likely to be attributed to the intercourse, and further sexual life is abruptly terminated.

Some patients are angry at having been singled out to suffer this affliction ('Why me?'); some grieve for what they have lost ('That's me finished'); and some are concerned about the effect of their illness on others ('My wife and I had been looking forward to our retirement'). Others assume that the stroke will go away like a cold in the head and that they will resume life where they left it off.

Physical examination

The description relates to the examination of the fully conscious patient once the condition has settled down. The patient is examined first in bed, then sitting and then standing, although this cannot always be conveniently done at one session. Patients fatigue easily, and find it difficult to cooperate in tests which require fine discrimination. Bedside tests are therefore brief and undemanding.

Examination in bed

The following observations supplement the standard neurological examination.

Deviation of the head at rest to the affected side may accompany disturbance of consciousness. Deviation of the head and eyes to the unaffected side indicates unilateral visual agnosia. The patient does not turn the head or eyes when approached or addressed from the affected side, although they may reply promptly to questions addressed to them from that side.

The immobile arm may be adducted and flexed at shoulder, elbow, wrist, thumb, and fingers, indicating spasticity; or it may lie limply at the side of across the chest, or under the body, suggesting proprioceptive loss. If the arm is weak it is supported by the examiner during testing, to avoid stressing the shoulder joint. Voluntary movement is tested by asking the patient to perform a meaningful action, e.g. 'Take this apple', rather than a meaningless one, e.g. 'Raise your arm'. The movement can sometimes be stimulated by mentioning an imaginary situation that is rich in associations, for example: 'It is a hot day and you have ordered a pint of ice-cold beer. There it is in your hand. Now let me see you lift it up and drink it.' Minor weakness is detected by asking the patient to raise both arms, and observing sagging of the weaker arm.

Extension of the hip, knee, and ankle suggests spasticity. This is confirmed if sudden abduction of the normal leg causes involuntary adduction of the affected leg. Voluntary movement is tested by supporting the leg against gravity. Minor degrees of weakness are tested by asking the patient to raise the extended limb from the bed.

Observation in a chair

Patients who sit upright without support have good postural tone. Those who slump have lost postural tone, because of the severity of the stroke, the weakening effect of accompanying illness, or prolonged immobility. Patients who slump to one side, leaving a triangular bare area on the back of the chair, have agnosia of proprioception on the side opposite to the triangle.

Standing

The patient stands once postural trunk tone is good and the affected lower limb is strong enough to take weight. The hands are clasped and the arms are held out in front of the trunk, bringing the centre of mass of the body forward and central. The outstretched arms are supported, and the affected leg is stabilized at the knee joint. Observation is made of the symmetry of the stance, any tendency to push the body towards the affected side, and the stability of the knee.

Tests of brain function

The following informative tests of brain function are carried out quickly at the bedside in all stroke patients, without fatigue or annoyance. The required equipment is a table, a felt-tip pen, two sheets of plain A4 paper, a newspaper, an apple, and an orange.

- Write your name and address.
- Read the newspaper.
- Describe a picture.

- Open the newspaper.
- Draw a house.
- Hand touch.
- Thumb finding.
- Apple–orange.
- Take off a cardigan.
- Put on a cardigan.

Write your name and address

A sheet of A4 paper is placed on the table in front of the patient with the long edge horizontal, and s/he is asked to write his or her name and address.

Patients with *apraxia* cannot hold the pen, or they hold it the wrong way round, or they move it about aimlessly in the air, or they draw circles on the paper or beside the paper. Patients with *agnosia* are puzzled by the pen, do not know what to do with it, explore various possibilities and may eventually put it in their mouth.

Patients with *right hemiplegia* may be unable to write; or may write with the left hand. They tend to print, rather than to use cursive script, possibly because printing is a simpler form of writing and is learned earlier in life. Some patients print individual letters scattered over the page.

Patients with *left hemianopia* commence writing as normal from the left margin, but turn their head to bring this into their visual range; whereas those with severe, unilateral, *left visual agnosia* commence in the middle of the page and write only in its right half. In cases of *incomplete* left visual agnosia the severity can be roughly gauged by observing how far along the page the writing starts.

The writing of patients with *left hemiplegia* and *balance disorders* often slopes up from left to right.

The writing of some patients is characterized by repetition of letters, words, ideas or the complete task, or of responses to previous instructions. This is a manifestation of perseveration, a general phenomenon associated with extensive brain lesions.

Read the newspaper

Patients are given a tabloid newspaper. They are asked to read the headline, and to read and summarize a short paragraph. Those with *aphasia* or *alexia* cannot read at all.

Patients with *hemianopia* turn the head towards the hemianopic side in order to bring the paper into the unaffected visual field, but then read normally. Those with *unilateral visual agnosia* hold the paper straight in front of them, but read only the words which are in the normal visual field, omitting the others, even although what they read makes no sense. Patients with *severe agnosia* omit the first half of every line, while those with lesser degrees omit only the first word or two, or hesitate before finding the beginning of the line, or fill in the missing word with a guess.

Some patients who are asked to read the headline select a relatively minor item in the lower half of the page. This failure to organize material in an hiearchical order is, like perseveration, a general manifestation of severe brain damage.

While the ability to summarize depends on previous intelligence, an unexpectedly poor summary of a short paragraph, or a mere repetition of what was read, imply general brain damage.

Describe a picture

A picture is selected in which there is something to describe in both halves, e.g. a picture of a bride and groom.

Patients with *unilateral visual agnosia* observe only what is in one half of the picture, and describe it without expressing its significance. For example, a wedding photograph of a bride and groom is described just as 'a lady'.

Patients with *partial bilateral visual agnosia* may read the caption instead of describing the picture. Those with extensive brain damage may fail to scan the picture, describe one detail like a dog carried by a guest, and announce that it is a picture of a dog.

Open the newspaper

This serves as a test of dexterity for those with minor motor or sensory loss. It also distinguishes between adaptability and perseveration.

Hand touch

The patient is touched lightly on the back of one hand, then the other, then both simultaneously. Patients with *hemianaesthesia* feel the touch on the unaffected but not on the affected side. When both sides are touched this is felt only on the unaffected side. Patients with *tactile agnosia* feel the touch on either hand; but when both are touched simultaneously this is felt only on the unaffected side.

Draw a house

The patient is given a fresh sheet of paper on which to draw a house. If they fail completely the examiner draws a stylized house and asks the patient to copy it. A normal drawing is two-dimensional and childish. It comprises walls, roof, windows, and door, often with the addition of a path and a chimney. Those with artistic training make a three-dimensional, detailed drawing.

Patients with *unilateral visual agnosia* omit the left-hand walls. Those with *visual agnosia* draw a house whose parts are not in proper relation to the whole, e.g. the windows are outside the walls, the chimney is beneath the roof. Patients with extensive brain damage draw a very small house or a very big one or one with many absent features.

Thumb finding

The examiner supports the patient's affected arm and asks him or her to grasp the thumb with the unaffected hand and then let it go. The examiner then covers the patient's eyes and moves the arm to a different position. The patient is asked to grasp the thumb again. Patients with *loss of proprioception* seek the thumb in the position where they last saw it, as they have not perceived the motion of the arm. Alert patients who have previously failed this test may reach for the shoulder and feel their way down to the thumb.

Apple–orange

The examiner conceals two large brightly coloured objects behind his or her back, one in each hand, and suddenly produces them in front of the patient. An apple and an orange are ideal for the demonstration of gross abnormality, but smaller objects, like a red and a blue pen, are better for the detection of lesser abnormalities.

Patients with *hemianopia* flick their gaze immediately towards the object in the intact visual field, then bring their eyes slowly back to the midline. They may then explore the other visual field.

Patients with *severe unilateral visual agnosia* look straight ahead. When they are asked what they see they mention only the object in the unaffected visual field, e.g. the apple. The examiner then interchanges the two objects in full view of the patient and asks again, 'What do you see?' Patients with severe unilateral visual agnosia say that they now see the orange, and express no curiosity about the fate of the apple.

Take off a cardigan

The patient is dressed in a loose-fitting, woollen cardigan and is instructed to take it off. Patients with *'operative' motor lesions* may pull the cardigan over the head, shake the unaffected arm out of the sleeve, and use the normal hand to pull the other sleeve over the affected arm. Those with motor lesions at 'administrative' level, i.e. *apraxia*, cannot reprogramme the synergy, and persist in making futile attempts to perform the now defunct programme which was successful before the stroke. Those with *agnosia* are unable to perceive the relationship between the body and the garment. Their activities are fragmented and apparently purposeless.

Put on a cardigan

The cardigan is placed across the patient's knees and instruction is given on how to put it on. Patients with *'operative' motor lesions* identify the different parts of the cardigan, arrange them conveniently, use the unaffected hand to guide the affected one through the correct sleeve, pull the sleeve over the arm and shoulder, catch the cardigan from behind, pull it over the unaffected shoulder, and put the normal hand through the sleeve.

Patients with *apraxia* attempt to thrust the affected arm into the correct sleeve, and fail because they make no allowance for the altered shape of the affected limb. They are puzzled what to do next, and usually try the same move again. They may put the unaffected arm into the wrong sleeve and persist in attempts to continue the sequence.

Patients with *agnosia* cannot identify sleeve holes, and cannot determine the relation of the parts of the cardigan to the whole, nor the relation of the garment to the body. They may make futile, disorganized attempts and then abandon the task.

Patients with severe, generalized brain damage make a series of short, stabbing movements which are of insufficient length or duration to achieve anything worthwhile.

The results of these rapid tests are summarized in the next box.

Rehabilitation

Rehabilitation of the stroke patient follows the same principles as were set out in Chapter 4, but with special challenges, which require careful assessment and exchange of information, and day-to-day coordination between members of the rehabilitation staff, patients, and relatives.

The techniques used are set out in appropriate texts. Those based on neurophysiological principles are the most popular. In elderly patients, technique is less important than clear definition of aims, continuous re-assessment, and consideration of the special factors which complicate the life of the stroke patient. Some of these are now discussed briefly.

Accompanying illness

Rehabilitation is more difficult in the presence of angina, heart failure, anaemia, arthritis, anaemia, and other painful and fatiguing conditions.

Complications

Pressure sores, urinary infection, incontinence, spasticity, contractures, subluxation of the shoulder, pulmonary aspiration, infections, falls, fractures, and epileptiform seizures delay rehabilitation and sap the patient's strength and will.

Learning

Learning to dress or walk with a paretic limb may necessitate unlearning and relearning, which are very much more difficult in an elderly person with a damaged brain than they are in a young person with an undamaged brain. Many patients seem to be unable to carry progress from one session to the next.

'Administrative' functions (see Chapter 13)

Stroke illness often affects those brain functions which are necessary for learning, especially memory and the ability to relate current percepts to past experience.

Neurological recovery

There is often considerable recovery in the early stages, as brain oedema subsides; but further recovery of function is slow and may be dispiriting.

Speech and language disorders

Inability to comprehend or to communicate effectively with the therapist is troublesome in dysphasic patients. Some patients with right-sided brain lesions show disconnection between words and actions; they say what they are supposed to do, but do something else.

Cognitive impairment

Multiple and extensive lesions impair the ability to understand the situation and to follow instructions.

Depression

Many patients despair of returning to anything approaching their former quality of life.

SUMMARY OF RAPID TEST RESULTS

1. Write your name and address

FINDING	SIGNIFICANCE
Unable to use pen	Apraxia
Does not understand pen	Agnosia
Cannot put pen to paper	Agnosia
Incomplete	Brain damage
Wrong address	Brain damage
Starts in middle of page	Left visual agnosia
Prints	Dysgraphia
Slopes up from left to right	Left hemiparesis

2. Read the newspaper

FINDING	SIGNIFICANCE
Unable to read	Aphasia, alexia, non-native
Occipital blindness	English, non-literate
Turns head to one side	Hemianopia
Omits beginning of line	Left visual agnosia

3. Describe a picture

FINDING	SIGNIFICANCE
Omits object on one side	Unilateral visual agnosia
Reads caption	Visual agnosia
Focuses on one detail	Left hemispher lesion

4. Open the newspaper

FINDING	SIGNIFICANCE
Unable to hold	Hemiparesis
Unable to turn page	Hemiparesis; hemianaesthesia
Perseveration	Brain damage

5. Hand touch

FINDING	SIGNIFICANCE
Does not feel on single touch	Hemianaesthesia
Feels on single but not on double touch	Tactile agnosia

6. Draw a house

FINDING	SIGNIFICANCE
Unable to use pen	Apraxia
Does not understand pen	Agnosia
Cannot put pen to paper	Agnosia
Omits left side	Unilateral visual agnosia
Faulty spatial relations	Visual agnosia
Very large, very small	Brain damage
Lacks details	Brain damage

7. Thumb finding

FINDING	SIGNIFICANCE
Fails to find thumb	Proprioceptive loss
Does not understand instructions	Left-hemisphere lesion

8. Apple–orange

FINDING	SIGNIFICANCE
Eyes move to side and return	Hemianopia
Eyes stay straight but object is not identified	Unilateral visual agnosia

9. Take off cardigan

FINDING	SIGNIFICANCE
Successful	No cortical involvement
Cannot learn sequence	Apraxia; agnosia
Futile, repetitive movements	Brain damage

10. Put on cardigan

FINDING	SIGNIFICANCE
Successful	Nor cortical involvement
Cannot learn sequence	Apraxia; agnosia
Futile, repetitive movements	Multi-infarct

Altered self-image

Patients are highly sensitive to their altered appearance and abilities, distorted language, facial paralysis, drooling of saliva, displacement of dentures, and especially an uncontrollable tendency to weep.

Inconsistency

Even in the best organized departments, patients tend to be handled inconsistently, which is bewildering and frustrating.

Good intentions

Among the less helpful attitudes are blaming patients for not trying; encouraging them with unjustified optimism; telling them that it is up to them; consoling them by telling them that others are worse off; and advising them that they will just have to learn to live with it.

The physical environment

All the physical inconveniences of hospital, home, and the exterior become complex and expensive problems.

Finance

There are many worrying expenses in connection with loss of work, travelling costs, adaptations to the house.

Change of roles

The disability of stroke diminishes the patient and changes his or her role in the family.

Notes and references

Clinical features

Among studies which emphasize the non-motor neurological features of stroke are those of Adams and Hurwitz (1963) and of Adams (1974) on perceptual and cognitive deficits in stroke patients who failed to recover; Ullman's (1962) description of behavioural changes resulting from the stroke patient's faulty perception of the world; the analysis by Fisher (1982) of disorientation for place; the demonstration by Smith *et al.* (1983) that loss of proprioception and neglect of space occurred in 40 per cent of stroke victims; and the description of patterns of cognitive loss by Wade *et al.* (1989). A guide to the less familiar manifestations of stroke illness is given by Pemental (1986).

Other clinical features of stroke illness which have received attention in the literature include swallowing difficulties (Selley 1985; Campbell-Taylor and Fisher 1987; Horner and Massey 1988); epileptiform seizures (Gupta *et al.* 1988); malalignment and dislocation of the shoulder joint (Najenson and Pikienly 1965; Fitzgerald-Finch and Gibson 1975); and depression, especially in right-sided lesions (Finset *et al.* 1989).

Neuropathology

Among pathological studies on the relationship between brain lesions and function are the work of Luria (1973) on World War II victims of head injury; the differentiation by Piercy *et al.* (1960) of the symptoms of apraxia; the attribution by Geschwind (1965) of the manifestations of focal brain disease to disconnection between functioning parts of the brain; and the demonstration by Kertesz and Ferro (1984) that apraxia is caused by lesions extending well beyond the parietal lobe.

Assessment

Wade and Collin (1988) made a case for the adoption as standard of the Barthel index (Mahoney and Barthel 1965). The score on this test corresponds with other assessment procedures; with measures of gait speed, stride length and postural stability (Pettmann *et al.* 1987); and with outcome after leaving hospital (Granger *et al.* 1988). One of its disadvantages is the 'ceiling effect' which renders it insensitive to the detection of continuing recovery (Skilbeck *et al.* 1983).

Rehabilitation

Brocklehurst *et al.* (1976) and Davies *et al.* (1989) found that only one-third to one-half of stroke patients in the community received therapy, while those who received most treatment were considered least likely to benefit. Wagenaar *et al.* (1990) used standardized assessments and single-case methods to compare the effectiveness of the Bobath (1978) and Brunnstrom (1956) approaches to rehabilitation, but found no significant difference between them. Edmans and Lincoln (1989) were unable to demonstrate success in treating the visual-perceptual deficits associated with stroke.

References

Adams, G. F. (1974). *Cerebrovascular disability and the ageing brain.* Churchill Livingstone, Edinburgh.

Adams, G. F. and Hurwitz, L. J. (1963). Mental barriers to recovery from stroke. *Lancet*, **ii**, 533–5.

Bobath, B. (1978). *Adult hemiplegia: evaluation and treatment.* Heinemann, London.

Borrie, M. J., Campbell, A. J., Caradoc-Davies, T. H., and Spears, G. F. S. (1986). Urinary incontinence after stroke: a prospective study. *Age and Ageing*, **15**, 177–81.

Brocklehurst, J. C., Andrews, K., Richards, B., and Laycock, P. J. (1976). How much physical therapy for patients with stroke? *British Medical Journal*, **1**, 1307–10.

Brunnstrom, S. (1956). Associated reactions of the upper extremity in adult patients with hemiplegia: an approach to training. *Physical Therapy Review*, **36**, 225–36.

Campbell-Taylor, I. and Fisher, R. H. (1987). The clinical case against tube feeding in palliative care of the elderly. *Journal of the American Geriatrics Society*, **35**, 110–4.

Davies, P., Bamford, J., and Warlow, C. (1989). Remedial therapy and functional recovery in a total population of first stroke patients. *International Disability Studies*, **11**, 40–4.

Edmans, J. A. and Lincoln, N. B. (1989). Treatment of visual perceptual deficits after stroke: four single case studies. *International Disability Studies*, **11**, 25–33.

Finset, A., Goffeng, L., Landry, N. I., and Haakonsen, M. (1989). Depressed mood and intrahemispheric location of lesions in right hemisphere stroke patients. *Scandinavian Journal of Rehabilitation Medicine*, **21**, 1–6.

Fisher, C. M. (1982). Disorientation for place. *Archives of Neurology*, **39**, 33–6.

Fitzgerald-Finch, O. P. and Gibson, I. I. J. M. (1975). Subluxation of the shoulder in hemiplegia. *Age and Ageing*, **4**, 16–18.

Geschwind, N. (1965). Disconnection syndromes in animals and man. *Brain*, **88**, 237–94; 585–644.

Granger, C. V., Hamilton, B. B., and Gresham, G. E. (1988). The stroke rehabilitation outcome study. Part 1: general description. *Archives of Physical Medicine and Rehabilitation*, **69**, 506–9.

Gupta, S. R., Naheedy, M., Elias, D., and Rubino, F. A. (1988). Post-infarction seizures: a clinical study. *Stroke*, **19**, 1477–81.

Horner, J. and Massey, E. W. (1988). Silent aspiration following stroke. *Neurology*, **38**, 317–19.

Kertesz, A. and Ferro, J. M. (1984). Lesion size and location in ideomotor apraxia. *Brain*, **107**, 921–33.

Luria, A. R. (1973). *The working brain: an introduction to neuropathology.* Pergamon, London.

Mahoney, F. L. and Barthel, D. W. (1965). Functional evaluation: the Barthel index. *Maryland State Medical Journal*, **14**, 61–5.

Najenson, T. and Pikienly, S. S. (1965). Malalignment of the glenohumeral joint following hemiplegia: a review of 500 cases. *Annals of Physical Medicine*, **8**, 96–106.

Pettmann, M. A., Linder, M. T., and Sepic, S. B. (1987). Relationships among walking performance, walking stability and functional assessment of the hemiplegic patient. *American Journal of Physical Medicine*, **66**, 77–90.

Piercy, M., Hecaen, H., and Ajuriaguerra, J. (1960). Constructional apraxia associated with unilateral cerebral lesions: left and right side compared. *Brain* **83**, 225–42.

Pimental, P. A. (1986). Alteration in communication: biopsychosocial aspects of aphasia, dysarthria and right hemisphere syndromes in the stroke patient. *Nursing Clinics of North America*, **21**, 321–37.

Selley, W. G. (1985). Swallowing difficulties in stroke patients: a new treatment. *Age and Ageing*, **14**, 361–5.

Shinton R. A., Gill, J. S., Melnick, S. C., Gupta, A. K., and Beevers, D. G. (1988). The frequency, characteristics and prognosis of epileptic seizures at the onset of stroke. *Journal of Neurology, Neurosurgery and Psychiatry*, **51**, 273–6.

Skilbeck, C. E., Wade, D. T., Langton Hewer, R., and Wood, V. A. (1983). Recovery after stroke. *Journal of Neurology, Neurosurgery and Psychiatry*, **46**, 5–8.

Smith, D. L., Akhtar, A. K., and Garraway, W. M. (1983). Proprioception and spatial neglect after stroke. *Age and Ageing*, **12**, 63–9.

Ullman, M. (1962). *Behavioural changes in patients following strokes*. Thomas, Springfield, IL.

Wade, D. T. and Collin, C. (1988). The Barthel index: a standard measure of physical disability? *International Disability Studies*, **10**, 64–7.

Wade, D. T., Skilbeck, C., and Langton Hewer, R. (1989). Selected cognitive losses after stroke: frequency, recovery and prognostic importance. *International Disability Studies*, **11**, 30–9.

Wagenaar, R. C. *et al.* (1990). The functional recovery of stroke: a comparison between neurodevelopmental treatment and the Brunnstrom method. *Scandinavian Journal of Rehabilitation Medicine*, **22**, 1–8.

Aphorisms

Of stroke and hemiplegia

- Stroke and hemiplegia are not synonymous. A stroke is an event in time, hemiplegia a manifestation in space.
- Not all hemiplegia is caused by stroke; not all stroke is manifested as hemiplegia.
- Looking on a stroke as 'just hemiplegia' is like looking on the United States as 'just New York'.
- Loss of movement is seen; loss of sensation is unseen.
- A stroke is a bomb in the brain. The disturbance which it causes spreads far beyond the site of the damage.

Right and left

- A right-sided stroke affects the right half of the body. A 'right sided CVA' is an ambiguous term and should be avoided.
- Anatomically there are two kidneys and one brain. Functionally there is one kidney and two brains. The kidney is divided for

convenience of packing. The two brains are united for convenience of communication.

- In the brain, as in marriage, dominance is replaced by specialization. The right and left brains, like husband and wife, each perform tasks which cannot be interchanged, and tasks which can be interchanged.
- The right brain is the space brain; the left brain is the speech brain.
- After a stroke the right brain does not know what the left brain is doing, and the left brain does not know what the right brain is saying. That is why left hemiplegics are optimistic and right hemiplegics are pessimistic.
- Left hemiplegics get home after a stroke because they can talk their way out of hospital. Right hemiplegics stay in after a stroke because they cannot talk their way out of hospital.

Of the clinical features of stroke

- A patient who loses proprioception is like a householder who is burgled. He does not know the value of his possession until he loses it.
- The hemianopic sees not and knows that he sees not; the visual agnosic sees not and knows not that he sees not.
- The apraxic does not and knows not that he does not; the agnosic knows not and knows not that he knows not.
- Emotional liability is a misnomer; emotional incontinence an ambiguity. When a stroke patient cries without being sad or laughs without being glad, he is suffering from emotional dissociation.
- In emotional dissociation the expression of emotion is dissociated from the experience of emotion.
- A painful hip is likely to be due to a lesion in the hip. A painful shoulder is likely to be due to a lesion in the shoulder. So why call it thalamic?

Of diagnosis and assessment

- Reflexes test the nervous system at its lowest level; speech tests the nervous system at its highest level. Students are failed for not examining the reflexes. Students are not failed for not examining speech.
- A tendon hammer is to a physician as a cane to a schoolmaster, a symbol of authority not an instrument of instruction.
- Eliciting tendon reflexes in stroke adds nothing to the assessment. No patient ever complained: 'Doctor, I have noticed that every time I stroke the sole of my foot my big toe goes up.'

- Proprioception is lost in the arm, not the finger. Proprioceptive loss should be demonstrated in the arm, not in the finger.
- A test is not necessarily good because it is costly nor bad because it is cheap. Yet given a choice between a CT scan and the 'Draw a man' test, most would go for the scan.
- An apple, an orange, a newspaper, a pencil and paper, and a cardigan tell more about the function of the nervous system than a CT scan and an EEG.
- The diagonal patient becomes the long-stay patient. A prognosis can be made from 10 m away by looking for the exposed triangle on the back of the chair.
- Watch the patient putting on his underpants and you will see the nervous system at work in all its glory.

Of rehabilitation

- A stroke victim facing rehabilitation is like a lone yachtsman crossing the Atlantic without map, compass or radio.
- A patient who lacks motivation lacks diagnosis.
- When the doctor considers the patient to be poorly motivated the patient probably considers the doctor to be poorly motivated.
- Circumduction is to gait as keloid is to a scar.
- After a stroke schoolteachers do worst because they are trying to get top marks for progress. Odd job men do best because they are satisfied with anything so long as it works.

15 *Aphasia*

Introduction

Man is distinguished from the brute creation by his ability to express thoughts and feelings in words. The loss of this ability deprives him of much of his humanity.

The study of aphasia, the loss of language ability, has fascinated neurologists and linguists; but has produced a complex and confusing terminology. In this chapter a simple scheme is presented to aid understanding and management of aphasic patients, which describes:

- what the patient means by what they say;
- what they understand of what is said to them;
- how to communicate effectively with the patient;
- how to tell others so that they too will understand.

The use of this method should be supplemented, modified, and altered by the professional assessment of a speech pathologist or speech therapist.

The scheme introduces three concepts:

- the levels of language;
- the channels of comprehension;
- the structuring of sentences.

The scheme is used to:

- assess how much language the patient uses when s/he speaks;
- determine how much language s/he comprehends;
- understand the structure of his or her speech.

The levels of language

Language can be thought of as being organized at five levels:

1. *New Speech*: newly formed and unique sentences.
2. *Old Speech*: previously formed and often used phrases and sentences like 'How are you?', 'Very well thank you'; 'It's a nice day'; 'It depends what you mean'; 'You can't teach an old dog new tricks'.
3. *Reactive speech*: single words or short phrases used in response to questions, like 'Yes', 'No', 'Please', 'Thanks', 'Fine', 'OK'.

4. *Emotive speech*: expressions of emotion without specific meaning, including expletives and swear words.
5. *Rote speech*: repetition of learned material that has been committed to heart, like the days of the week, the months of the year, the numbers from 1 to 10, poetry and prayers.

These comprise the Language Ladder (Fig. 15.1). There is no sharp delineation between the steps on the Ladder, and there is a wide range of complexity within each step.

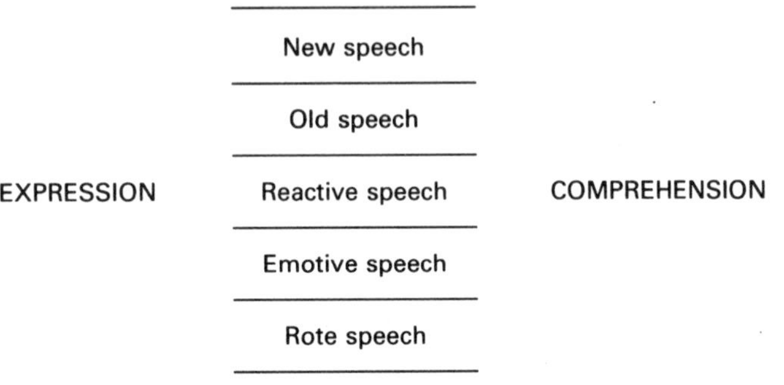

Fig. 15.1 The language ladder.

Normal conversations are conducted and understood in a mixture of these, with New speech prevailing. Aphasia is descent of the Language Ladder. Aphasic patients use and understand little New speech. They work mostly with Old, Reactive, and Emotive speech, while some use only Rote speech.

In addition the speech of the aphasic patient contains abnormal language. This will be considered in the section on The structure of sentences.

The Language Ladder is used to describe the levels of the patient's expression and comprehension of speech. These are tested together, but, for convenience of presentation they are treated separately.

Assessment of language

- *New speech* is tested by open-ended questions, such as 'Tell me about your work', or 'What do you enjoy doing?'
- *Old speech* is tested by questions such as 'How are you?' or 'What do you think of the weather?'
- *Reactive speech* is tested by closed questions like 'Would you like a drink?'

- *Emotive speech* occurs spontaneously or as an inappropriate reply to a question.
- *Rote speech* is tested by asking the patient to recite the days of the week, or the numbers from 1 to 10, or a poem. Clues may be used to help, such as the first one or two members of the series or lines of the poem.

Severely aphasic patients may retain the ability to sing the words of a song, while being unable to speak the same words.

Some examples of speech-ladder testing are given in the next boxes.

Example 1

Q. Hello, how are you?
A. Fine. (**Reactive speech**)

The patient used Reactive speech, but may be able to use Old and New speech.

Q. What is your address?
A. 25 Gardiner Street. (**Rote speech**)

The given address may or may not be correct.

Q. Tell me what family you have.
A. Three girls and a boy. (**Old speech**)

The answer sounds correct, and the patient appears to have understood the question. The language used in the reply lies on the border between Old and New speech. S/he did not use a grammatical sentence.

Q. What did you have for breakfast?
A. I had eggs . . . I had eggs and eggs . . . I don't know. (**Old speech**)

Again the patient appeared to understand the question, but had difficulty in formulating the answer, and used abnormal language in Old speech.

Conclusion

From the responses to these few questions it can be tentatively inferred that the patient functions at the level of Old speech in relation to expressive language.

Example 2

Q. Hello, how are you?
A. Fine. (**Reactive speech**)

The patient used Reactive speech, but may be able to use Old and New speech.

Q. What is your address?
A. Fine. (**Reactive speech**)

This answer is an inappropriate repetition of the previous answer.

Q. Well, tell me your name.
A. [No reply]

The patient appears to have only a low level of language at his command.

Q. Are you married?
A. Yes. (**Reactive speech**)

He has given a second example of his use of Reactive speech.

Q. Tell me your wife's name
A. [No reply]

He has no access to a familiar name.

Q. Now I would like you to count from 1 to 10. Let's start together. One . . . two . . .

A. Three . . . four . . . four . . . (**Rote speech**)

Conclusion
It can be tentatively inferred that the patient is to be placed on the 'Emotive speech' rung of the Speech Ladder, and has little or no access to higher levels of language.

The channels of comprehension

Language barriers to comprehension

The comprehension of spoken language is affected by accent, vocabulary, memory, speed, 'spectrum', and fatigue.

The speech signals conveyed by different *accents* are differentiated by the intact brain to give meaning. For example the sound [BOON] might mean a blessing if spoken by an Oxford clergyman; but a Yorkshire housewife would be referring to something to eat for tea. The ability to relate variant sound to meaning is impaired in aphasia. Consequently, aphasic patients comprehend the familiar accents of members of their own family better than those of doctors and nurses.

Aphasic patients have difficulty in dealing with unfamiliar *vocabulary* or sentence structure. They therefore understand the language of their own family better than they do that of the hospital staff. Aphasic patients may have impairment of short-term *memory*. A long sentence is not understood if the first words are forgotten before the final words are pronounced. This occurs especially if the sentence includes the word 'or'. The patient who is asked 'Would you like to go out, or would you prefer to sit here and watch television?' may answer 'Yes', indicating that he or she has recalled only the second half of the sentence.

The *speed* of processing of spoken language is reduced in aphasic patients. Their ability to understand is dependent on their having sufficient time to process what they hear. So speech should be slowed, but it should not be so slow that the patient is in danger of forgetting the beginning of the sentence before the end is reached.

Every word has a profile or *spectrum* of sound. The words 'television' and 'spectacles' contain several syllables, do not sound like any other words, convey strong visual images, and relate to familiar objects. They might be called 'high profile' words, and are easily understood. By contrast the words 'give' and 'show' are short, undistinctive, and have no specific visual associations. They are 'low profile' words. When aphasic patients are asked 'Give me your spectacles', they may respond appropriately, or they may do something else with their spectacles, indicating that they have understood the high-profile word but not the low-profile ones.

Fatigue sets in readily in aphasic patients, and performance falls off, or they suffer a 'catastrophic' reaction and burst into tears if they are overloaded. Testing should be kept short, and the maximum value must be made of every question.

Non-language aids to comprehension

Spoken language is characterized or accompanied by four factors which aid its comprehension: redundancy, music, gesture, and context. The ability to make use of these aids may be fully or partially preserved when language comprehension is impaired. This is exemplified by the sentence. 'Please will you tell me the way to the post office.'

Language uses *redundant* phrases; it is possible to understand the sentence without understanding the first three words.

Spoken language is *musical*. It is accompanied by changes of pitch, tone, speed, rhythm, and volume, which impart meaning. The sample sentence can be understood as being a polite or an exasperated request according to the speech music which accompanies it.

Spoken language is accompanied by *gesture*, facial expression and other forms of body language. The sample sentence can be made comprehensible by gesture alone.

Spoken language takes place in a spatial and social *context*. The sample sentence can be comprehended fully by observing that the person who asks the question looks like a stranger in town and is carrying a bunch of postcards.

In conversing with an aphasic patient, maximum use is made of redundancy, music, gesture, and context to aid the patient's understanding. In testing the patient's ability to comprehend spoken language, minimum use is made of music, gesture, and context in order to determine how much the patient has understood the words themselves, and how much has been inferred from the other factors.

An example is asking the patient to put a spoon on a saucer. This can be worded: 'Now I am going to ask you please to pick up the spoon and put it down again on top of the saucer' using much redundant material; or simply: 'Put the spoon on the saucer.' The request can be made with much or little speech music. The questioner can point to or look at the objects to be placed together; or ignore them. The context can be helpful, when the patient is sitting at a table where a cup and saucer are placed on one another with a spoon nearby; or it can be unhelpful, when the table is covered with a random selection of objects.

Testing comprehension

The following points are to be borne in mind in testing comprehension.

- Every answer should advance understanding of the patient's problem.
- Questions should be short and should not contain the word 'or'.
- A different answer may be given to the same question at different times, because of fluctuations of concentration.
- A wrong answer may be given to a question which has been understood, because the patient lacks the ability to express the correct answer.
- A correct answer may be given to a question which has not been understood, because of the clues afforded by redundancy, speech music, body language, and context.
- A correct answer may be given to a question which has not been fully understood, because one prominent word has been understood and the rest has been guessed.

- A correct answer to a closed question does not necessarily mean that the question has been fully understood. This is challenged by posing an opposite question. For example, if a patient answers 'Yes' to the question 'Are you married?', he should be asked, 'Are you single?' If he gives the correct answer to 'What town do you live in', he should be asked, 'Do you live in . . . (a different town)?'
- The closer the questioning is to everyday matters, the better comprehension is likely to be.
- Finer levels of comprehension are tested by substituting inappropriate, unexpected, or nonsense words in a sentence which has been apparently correctly answered.

Boxed examples will now be given of identifying the patient's position on the comprehension side of the Language Ladder.

Example 1B

Q. Tell me about your family.
A. Lovely family.

She understood the word 'family', and may have understood the rest of the question. She does not appear to be able to use New speech.

Q. What family do you have?
A. Three boys and a girl.

She understood the whole question and replied appropriately in Old speech.

Q. Is your oldest married?
A. Yes.

This closed question provides no information on the patient's comprehension.

Q. Is your oldest single?
A. Yes.

The question contains some ambiguity. The child could be divorced or widowed and living alone, which might be interpreted as 'single'. But the answer suggests that the patient is having difficulty in comprehension, and a change of direction is advised.

Q. How would you make a cup of tea?
A. No thank you.

Example 1B *contd.*
The patient may have understood the words 'cup of tea'; but may have failed to distinguish between two low-profile words, 'make' and the more expected 'like'.

Interpretation
The patient appears to understand Old speech, but not New speech, because of difficulty in differentiating similar sounding words and phrases. She might be able to make use of contextual clues.

Example 2
Q. Tell me about your family.
A. [No reply]

She may have understood but be unable to use New speech.

Q. What is your address?
A. [No reply]

She may have understood but be unable to use Old speech.

Q. Is your address 25 Gardiner Street?
A. Yes.

She may have understood. The reply is in Reactive speech and may be purely random.

Q. Is your address 45 Gardiner Street?
A. Yes.

She failed to distinguish between two similar numbers.

Q. Is your address 25 Thomson Street?
A. Yes.

She failed to distinguish between two similar names.

Q. Is your address 25 Buckingham Palace?
A. No.

She can use a negative. She can recognize wholly inapplicable information.

Example 2 *contd.*

Q. Is your address green?
A. No.

She distinguished between 'address' and 'dress' because the context favoured 'dress'.

Q. What colour is your address?
A. Green.

She failed to distinguish between 'address' and 'dress' because the context favoured 'dress'.

Interpretation
She does not understand New speech and probably not Old speech. Further testing is necessary to determine how much she comprehends of gesture and context.

Testing comprehension by performance

In patients with no reactive speech, comprehension of language can be tested by observing responses to instructions, assuming that the patient is capable of making the movement. Instructions may be given with and without the aid of speech music, gesture, and context, in order to determine how much these contribute to comprehension. The instruction should be a sensible one. Touching the right ear with the left hand and putting out the tongue are not part of normal behaviour and not appropriate tests of comprehension.

Prepositions (words that show the relation between one noun (see below) and another) used in these tests have a low profile, and are easily misunderstood. The patient who responds correctly to the instruction 'Take off your spectacles', may fail to understand the instruction 'Take the pen and put it under the plate' because this requires understanding of nouns (words that name), verbs (words of doing or being), and prepositions; because the instruction has two elements; and because the action is not part of normal behaviour.

In testing comprehension by instruction, progressively more complex tasks are offered. The number of items of information may be increased from 'Touch my hand' or 'Take off your spectacles' to 'Take my pencil and put it on the table between the knife and the tumbler.'

Comprehension of the names of objects is tested by arranging a knife, a comb, a key ring, a pair of scissors, or similar objects and repeatedly asking the patient to pick up one of them.

Summary of testing comprehension

At the end of the test the doctor will know:

- the level of language which the patient comprehends;
- the level of language which the patient does not comprehend;
- the comprehension of language in a social context;
- the comprehension of gesture and intonation.

The structuring of sentences

Normal language can be conceived as being built up in four stages:

- from sounds to words;
- from words to phrases;
- from phrases to sentences;
- from sentences to language.

The language of aphasic patients may contain errors at any or all of these stages.

From sounds to words

An ordinary word is composed of vowel and consonant sounds built up in a recognizable pattern. The speech of aphasic patients may contain

- wrong words;
- abnormal words
- a sound on its own.

Wrong words may sound like the intended word but have a different meaning, e.g. chair for hair; or it may have a similar meaning to the intended word but a different sound, e.g. chair for table; or it may have no connection with the intended word, e.g. glum for chair.

Abnormal words are built up out of the same basic sounds, but in random order, so that the word is not recognizable and has no meaning, e.g. *crof* or *spak*.

Patients may utter a *speech sound* which has no meaning. They may repeat this over and over again on its own, or insert it into their speech.

From words to phrases

In normal speech, words are assembled according to the rules of grammar and word order, to convey meaning. Phrases differ in length and richness. Prepositions link the words.

The phrases used by aphasic patients may consist of only one word, which may be a swear word; or they may be fragmented with no real meaning, like 'I want to . . .'; or they may ignore the rules of grammar and word order, like 'I said he . . .' or 'the food . . . not nice'. Prepositions may be omitted or used wrongly, e.g. 'I put bottle bed'. It may be possible to guess the intended meaning, or this may be obscure, e.g. 'She said give take I don't know'.

The patient may repeat the same word again and again, e.g. 'I said don't don't' or the phrase may consist of a random collection of words with no apparent connection, e.g. 'There was no total I say total France French'.

From phrases to sentences

A normal sentence consists of one, two, or more connected phrases which develop a complex meaning.

The sentences used by aphasic patients may contain very few words, perhaps only one unfinished phrase, e.g. 'Give me . . .'; or there may be breaks between phrases, e.g. 'I want to . . . I need . . . she said . . .' Or the sentences may consist of phrases which are assembled at random and lack meaning; or meaningful and meaningless phrases may occur together, e.g. 'I told you so but table went table police no I don't mean that.'

From sentences to spoken language

The spoken language of aphasic patients may lack speech music and gesture; or these elements may be exaggerated or inappropriate, expressing frustration, apathy, indifference, anxiety, excitement, or depression.

Clinical patterns

While any or all of these types of error may occur in the speech of aphasic patients, they tend to be grouped into patterns, as follows.

- The patient speaks slowly and with difficulty, with few words and long pauses between. There are errors in words, in building words into phrases and phrases into sentences. There is little variation in speed, pitch, and tone, and much irrelevant movement of the face and body. This corresponds to Broca's aphasia.
- The patient speaks rapidly and with much expression, with many errors in words, and use of meaningless phrases and sentences. This corresponds to Wernicke's aphasia.
- There are errors in certain words, especially those for names of objects, but phrase and sentence structure are normal. This corresponds to anomic aphasia.

- There is very little spontaneous language, only grunts and non-speech sounds. This corresponds to global aphasia.

Classification of aphasia

The practical aim of classification is to separate patients who need to be managed in different ways. The classical classifications distinguish eight types, which is too many for practical purposes. The separation into 'motor and sensory' or 'expressive and receptive' is inappropriate to aphasia in stroke, where expressive and receptive difficulties coexist. The distinction between 'fluent' and 'non-fluent' types is easier and helps management; but one-third of aphasics do not fit into either category.

Non-fluent patients produce few words with effort. Their phrases are brief and telegrammatic; they lack rhythm and speech music; but they convey meaning. Patients know what they are trying to say and make great efforts to do so. They have difficulty in understanding others.

In fluent dysphasia, which is less common, patients have no difficulty in producing sentences of normal length and with full expression; but their meaning is obscure. Their speech is full of jargon, abnormal words, circumlocution, and repetition. They seem convinced that others understand them; but they do not understand others.

Agraphia and alexia

Agraphia, loss of the ability to write, and alexia, loss of the ability to read, are equivalent in written language to loss of expression and comprehension of spoken language. They often accompany aphasia. Account has to be taken of the patient's ability to read and write before the stroke. Testing is difficult when patients cannot speak or fully understand spoken language, and the help of a specialist may be needed; but much can be learned by asking the patient to read from a newspaper, point to words in the newspaper, write a name and address, and copy from the newspaper.

Dysarthric dysphasia

Dysarthria is usually described as a disturbance of speaking rather than of language, resulting from a lesion in the brain-stem, cerebellum, or extrapyramidal system. However, dysarthria may accompany dysphasia as a consequence of a lesion which straddles the contiguous brain areas where thoughts are turned into words and where words are produced as sounds.

Psychosocial aspects

Aphasia is sometimes accompanied by euphoria or '*belle indifference*', sometimes by uninhibited laughter, more often by uninhibited tears, and very often by intense frustration at the inability to communicate and at the great gulf that opens up between the patient and the familiar world.

Patients describe their sense of being a non-person when others walk past their beds, or walk away from a conversation that has become stuck; or speak to one another in their presence as though they were not there. They talk of the embarrassment of suddenly bursting into tears for no reason, and the constant anxiety that this might recur. They speak of the difficulties of sustaining a telephone conversation when they cannot see the effects which their words are having, and when they cannot recognize other people's words which have been distorted by the instrument.

This anguish can only be guessed at when patients are too aphasic to describe their feelings. It is salutary to speak to recovered patients and to their spouses about their perceptions at the time when they were cut off on the ice floe of aphasia, and left to drift far from all contact with the familiar world.

The course of aphasia in stroke

The aphasic patient is at risk of becoming constipated, dehydrated, incontinent, frustrated, and depressed.

Physical rehabilitation may be adversely affected by the aphasic stroke patient's isolation, inability to comprehend verbal instructions, failure to gain encouragement from conversation and sharing of experience with other patients, and difficulty in attracting help.

The language disorder shares the tendency of the physical manifestations of stroke to spontaneous recovery. In about one-quarter of patients the recovery is complete within the first three months. Other patients may not show any sign of improvement initially, but may begin to recover for up to six months after onset, and continue to improve over a substantial time. However, at least one-quarter of survivors show no true recovery at all, or reach a plateau where further recovery ceases, leaving severe residual disability.

Management of aphasia

Patients should be referred to a speech therapist for expert assessment and treatment. Hospital staff and those who live with the patient should be instructed by the speech therapist how to communicate with the patient.

The many approaches to therapy share the common ground of giving patients confidence; treating them as adults; stimulating all their senses; and never thwarting or humiliating them. A common misconception is to think of these patients as pupils who are learning a foreign language, rather than as adults who are trying to recover their own. The object is not accuracy but communication. The patient who asks for help to find the right words is given it; and the one who prefers to find the word for himself or herself is allowed to do so.

Terminating treatment on the grounds that no further benefit can be expected upsets patients and relatives; but individual treatment should not be unnecessarily prolonged. Treatment should be continud in a group. The Volunteer Stroke Scheme in the United Kingdom, and similar organizations in other countries, have demonstrated that dysphasic patients have more language ability than they use. Participation in a group with similar disabilities who play games and laugh together may bring this latent capability to the surface.

Notes and references

Classification

Reviews of the literature on aphasia are given by Benson (1979), Rose (1984) and Albert and Helm-Estabrooks (1988). A clear simple exposition is given by Gravell (1988), while Pimental (1985) adds a guide to approaching the patient and a check list. Enderby and Langton Hewer (1985) cover comprehensively the nature, diagnosis, and management of dysphasia, dyspraxia, and dysarthria.

Benson (1979) used the 'classical' classification into eight types, viz, Broca, Wernicke, conduction, transcortical sensory, transcortical motor, transcortical mixed, global, and anomic. Kertesz and Phipps (1977) provided a mathematical justification for a similar classification. Luria and Hutton (1977), whose work was undertaken originally on patients with focal traumatic lesions of the brain, placed emphasis on the motor and proprioceptive disturbances which affect control of the speech muscles; and related these to lesions in specific brain areas.

Geschwind (1971) attempted to simplify the problems of the clinician by reducing the classification to two categories, fluent and non-fluent, but admitted that many cases overlapped. According to Rose (1984), non-fluent patients produce their words with effort, and generally use fewer than 10 words per minute. Their phrases are brief and telegrammatic; they are dysrhythmic and lack speech music; but they convey meaning. Fluent dysphasics have no difficulty in producing sentences of normal length and with full expression, but their syntax is abnormal and their meaning is obscure.

They indulge in circumlocution, paraphasia, and jargon. Both types have difficulty in comprehension. Fluent dysphasia is only half as common as the non-fluent type.

Neuropathology

Huber *et al.* (1984) related the various subtypes of aphasia to the vascular pathology, and challenged the association between aphasia and involvement of the left frontotemporoparietal area. They showed that the lesion associated with aphasia can be large or small, anterior or posterior, cortical or subcortical, single or multiple; while in one-fifth of cases with 'amnesic' aphasia there does not seem to be any lesion at all.

Papanicolaou (1988) provided evidence that, during recovery from aphasia, there is increased activation of the right hemisphere.

Assessment

Bedside testing can provide a lot of information in a short time; but this may be misleading, and should be supplemented by formal tests (Rose 1984).

Treatment

After a remarkable experience in working with a young victim of subarachnoid haemorrhage (Griffith 1970, 1975), Griffith and Miller (1980) used volunteers, under the guidance of speech therapists, to bring back hope and stimulation to aphasic stroke victims. The results were as good as those gained by professional speech therapists (David *et al.* 1982).

Prognosis

Rose (1984) showed that one-quarter of patients with aphasia after stroke recovered completely within three months; and one-quarter remained severely affected after one year. Williams *et al.* (1977) found that stroke patients with pure dysphasia showed little benefit in the first three months after stroke, but subsequently improved; patients with dysphasia plus articulatory dyspraxia continued to improve for more than six months; and those with dysarthria alone did not benefit from treatment after three months.

Dementia

The language disorder in dementia is distinguished in a number of ways from that occurring after stroke. Aphasics seem to have less competence than early dements to retain in their mind more than two pieces of information, so that they cannot, for example, point to three separate items in a picture (Code and Lodge 1987). Dements, unlike stroke patients, may retain reading ability.

Parkinsonism

In a study of patients with advanced parkinsonism, Streifler and Hofman (1984) noted defective articulation of consonants, hypernasality, monotony, decay of tone, elongation of pauses, and palilalia. Parkinsonian patients have a disturbance of prosody or 'speech music', the quality which gives emotional expression to speech, and have difficulty in grasping the significance of the prosodic component of the speech of others (Scott and Caird 1983).

References

Albert, M. L. and Helm-Estabrooks, N. (1988). Diagnosis and treatment of aphasia. *Journal of the American Medical Association*, **259**, 1043–7; 1205–10.

Benson, D. F. (1979). *Aphasia, alexia and agraphia*. Churchill Livingstone, Edinburgh.

Code, C. and Lodge, B. (1987). Language in dementia of recent referral. *Age and Ageing*, **16**, 366–72.

David, R., Enderby, P., and Bainton, D. (1982). Treatment of acquired aphasia: speech therapists and volunteers compared. *Journal of Neurology, Neurosurgery and Psychiatry*, **45**, 952–61.

Enderby, P. and Langton Hewer, R. (1985). The context and management of acquired speech and language handicap. In *Recent advances in geriatric medicine*, Vol. 3, (ed. B. Isaacs). Churchill Livingstone, Edinburgh.

Geschwind, N. (1971). Aphasia. *New England Journal of Medicine*, **284**, 654–6.

Gravell, R. (1988). *Communication problems in elderly people: practical approaches to management*. Croom Helm, London.

Griffith, V. E. (1970). *A Stroke in the Family*. Penguin, London.

Griffith, V. E. (1975). Volunteer scheme for dysphasia and allied problems in stroke patients. *British Medical Journal*, **2**, 633–5.

Griffith, V. E. and Miller, C. (1980). Volunteer stroke scheme for dysphasic patients with stroke. *British Medical Journal*, **281**, 1605–8.

Huber, W., Poeck, K., and Willmes, K. (1984). The Aachen aphasia test. *Advances in Neurology*, **42**, 291–303.

Kertesz, A. and Phipps, J. B. (1977). Numerical taxonomy of aphasia. *Brain and Language*, **4**, 1–10.

Luria, A. R. and Hutton, T. (1977). A modern assessment of the basic forms of aphasia. *Brain and Language*, **4**, 129–51.

Papanicolaou, A. C. (1988). Evidence of right hemisphere involvement in recovery from aphasia. *Archives of Neurology*, **45**, 1025–9.

Pimental, P. A. (1986). Alteration in communication: biopsychosocial aspects of aphasia, dysarthria and right hemisphere syndromes in the stroke patient. *Nursing Clinics of North America*, **21**, 321–37.

Rose, S. and Caird, F. I. (1983). Speech therapy for Parkinson's disease. *Journal of Neurology, Neurosurgery and Psychiatry*, **46**, 140–4.
Streifler, M. and Hofman, S. (1984). Disorders of verbal expression in Parkinsonism. *Advances in Neurology*, **40**, 385–93.
Williams, B. O., Walker, S. A. and Dall, J. L. C. (1977). Who needs speech therapy? *Age and Ageing*, **6**, 96–103.

Aphorisms

Of aphasia

- Stroke is a bomb in the brain; aphasia is a knife in the cortex.
- Hemiplegia prevents walking from A to B. Aphasia prevents talking from B to A.
- To acquire aphasia is to arrive without a return ticket in a country where no English is spoken.
- Language is the highest human function. Loss of language is the deepest human dysfunction.
- Loss causes grief, anger, bewilderment, resentment. Aphasia causes grief, anger, bewilderment, resentment. When you find a battered wife, look for an aphasic husband.
- Aphasia is never expressive without being receptive and never receptive without being expressive.
- Speech is a piano and language is the score. When the strings of the piano are broken that is dysarthria; but when the score is damaged that is dysphasia.

Of aphasics

- Dr Samuel Johnson suffered aphasia and took alcohol, as he had observed that it stimulated loquaciousness. It had the desired effect.
- A recovered aphasic said: 'My worst moment was when two nurses made my bed and talked to each other as if I wasn't there.'
- To walk past the bed of an aphasic denies his humanity and diminishes one's own. To walk away from an aphasic is to sentence him to solitary confinement.
- It is a cruel jest of Nature to rob a man of language and leave him with only a swear word. It is a cruel action of Man to find this amusing.

Of management

- The silent aphasic is silent because he is inhibited not because he is aphasic.

- Hush in the classroom, hubbub in the playground: there lies the secret of the treatment of aphasia.
- Language is learned in social contexts. Lost language is relearned in social contexts.
- Work at restoring language and the gains are slow and small. Work at removing inhibition and the gains are swift and large.
- Correcting error inhibits; ignoring error disinhibits.
- The best disinhibitor of a dysphasic is another dysphasic. Together they cure one another.

16 *Visual impairment*

Introduction

Every patient who is received into a geriatric service may have some degree of visual impairment. This chapter gives guidance to the non-specialist on the information which should be sought about visual function in elderly patients.

Impairment of visual function

Visual function is impaired in late life by physiological and pathological changes in the lens, the iris, the retina, and the visual pathways in the brain. The diseases which commonly underlie these changes include refractive errors, cataract, glaucoma, hypertensive and diabetic retinopathy, macular degeneration, cerebral infarction, brain tumour, and Alzheimer's disease. Their effects include the following.

- Changes in the lens:
  increased rigidity;
  diminished translucency;
  cataract formation.
- Changes in the iris:
  diminution in pupil size;
  increase in intraocular tension.
- Changes in the retina:
  degeneration of the optic nerve;
  increased vascularization of the retina;
  retinal haemorrhages and exudates;
  new vessel formation;
  degeneration of the macula and peripheral retina;
  detachment of the retina.
- Neurological lesions:
  homonymous hemianopia;
  bitemporal hemianopia;
  other visual field defects;
  unilateral visual agnosia;
  visuospatial agnosia;
  occipital blindness;
  transient blindness.

Focal and ambient vision

Vision has two components which act in different ways, which are effected through different parts of the visual system, and which are variously impaired by disease.

Focal vision determines what is seen when an object is consciously looked at. It answers the question, 'What?' Focal vision is impaired by diseases of the lens, the media, and the retina.

Ambient vision subconsciously monitors the visual stimuli that flow into the eyes during motion, and facilitate location in space. It answers the question, 'Where?' Ambient vision is impaired by disease of visual pathways in the brain.

Effects of disease on focal vision

The effects of pathological changes on focal vision may include:

- alteration in near point;
- loss of visual acuity;
- increased dependence on level of illumination;
- reduction of visual field;
- distortion of visual image;
- intolerance of dazzle;
- difficulty in distinguishing figure from ground.

The three major causes of impaired focal vision in old age are macular degeneration, cataract, and glaucoma.

Macular degeneration

This is first noticed when small print becomes diffiult to read, and when larger print becomes blurred. Later, visual impairment affects household tasks, such as reading the controls on the cooker, observing steam from the kettle, noting the level of fluid in a cup. Domestic accidents are common, and cooking may have to be abandoned. Reading and watching television become increasingly difficult, and faces cannot be recognized. Mobility is retained at first, and the subject can even continue to drive; but night-time driving is first abandoned, then driving by day; and finally the patient becomes confined to the house.

Cataract

Cataract is so called because at the height of the disease vision is so blurred that it is as if the patient were looking through a waterfall. A more modern description is that the visual scene resembles that observed when the first drops of rain are smeared over a dirty windscreen. Bright light falling on the eyes is diffracted by the altered lens structure and the patient is dazzled. The patient holds a hand up to the eye or wears a peaked cap. Driving at night becomes dangerous because of the dazzle of oncoming headlights. If the cataract is central, faces are blurred and reading is difficult, but mobility is maintained.

Glaucoma

Glaucoma of the open-angle variety is very insidious. Its first manifestation may be a central scotoma, which may be unnoticed by the patient or described as blurring. Loss of peripheral vision proceeds slowly, and is unconsciously compensated for by increased head movement to scan the environment. Tunnel vision occurs late in the disease. Eyedrops for lowering intraocular tension may constrict the pupil and reduce the amount of light reaching the retina, aggravating the visual loss; but modern drugs do not have this undesirable effect.

Effects of disease on ambient vision

Severe impairment of ambient vision is associated with lesions involving the visual pathways and the occipital cortex. Lesser degrees of impairment may be due to central slowing or minimal cellular loss; or may be a feature of normal ageing. The effects of impaired ambient vision include:

- altered depth perception;
- impaired perception of distance;
- impaired discrimination of peripheral signals;
- impaired judgement of speed of approaching bodies.

Most of these changes occur gradually and are unnoticed by the patient until an accident draws them to attention.

Patients with impaired ambient vision cannot monitor the rapidly moving and changing visual scene when they walk in a busy street. They cannot make accurate judgements of the speed and distance of approaching persons or vehicles. People seem to loom up unexpectedly or to bump into them inexplicably. Patients may step out in front of a moving car. When they describe motor vehicle accidents they speak of the other car as 'having appeared out of nowhere'.

An error in depth perception may lead to a fall, when for example a patient believes a wall to be nearer than it is and attempts to avoid bumping into it.

Patients become frightened by the mismatch between reality and their perceived visual world. At first they compensate by slowing down, shortening the step, and stooping. Later they abandon crossing the street unless they are accompanied by a companion; or they stop going out. In the house, errors of depth and distance perception may cause falls, for example by 'missing' a chair or a step, or bumping into a piece of furniture or an animal.

Few patients are aware that they have impaired depth or distance perception. They may believe that they have normal eyesight because their focal vision is unaffected. Tests for ambient vision are not readily available. Impairment of ambient vision has to be inferred from the history, and should be sought in all patients who have fallen or who have abandoned going out alone for no obvious or sufficient reason.

Neurological lesions

Patients with homonymous hemianopia are rarely aware of visual field loss, for which they have compensated by using head movements. British patients with right hemianopia and European or American ones with left hemianopia are in danger on crossing the street from vehicles approaching from their hemianopic side. Homonymous hemianopia may be due to stroke, brain tumour, and other brain lesions. Bitemporal hemianopia may indicate pituitary tumour. 'Amaurosis fugax', or transient loss of vision in one eye, is due to momentary occlusion of the retinal artery by thrombotic material from the ipsilateral carotid artery. It may presage a major stroke on the opposite side. Unilateral and bilateral visual agnosia ('occipital blindness') were described in Chapter 14.

History taking

The routine medical history elicits the presence of diabetes and hypertension, which are associated with cataract, glaucoma, and retinopathy. The following questions are useful in the assessment of visual function.

- Do you have spectacles?
- For how long have you had them?
- For what disability do you require spectacles?
- For what purposes do you use your spectacles?
- How often do you have your eyes tested?
- When were they last tested?
- What was the outcome of that test?
- Have you ever been to an ophthalmologist?

- For what purpose?
- Do you have eyedrops?
- For what purpose do you have them?
- How do you use them?
- What changes have you noticed in your vision?
- Is there anything that you used to be able to do that you have had to stop doing because of your eyesight? Such as:
  driving;
  going out at night;
  crossing the street;
  cooking;
  reading;
  watching television;
  dressing.
- Do you have difficulty in reading:
  labels;
  small print in books and newspapers;
  large print books;
  newspaper headlines?
- Do you have difficulty in seeing to:
  thread needles;
  sew;
  knit, crochet;
  use a screwdriver;
  put an electric plug in a socket;
  put a key in the lock?
- Can you follow what is on the television?
- Are you troubled by glare from:
  sunlight;
  ordinary daylight;
  motor car headlights?
- Do you wear:
  a peaked cap;
  other brimmed hat;
  an eyeshade;
  dark glasses?
- Do you drive, and if so have you changed your driving habits because of your vision in any of the following ways:
  don't drive at night;
  don't drive far;
  drive more slowly;
  don't drive on motorways?

- Have you changed your habits as a pedestrian because of your vision in any of the following ways:
  don't go out at night;
  don't go out alone;
  don't try to cross busy streets;
  don't go into crowds?
- Have you been involved as a driver or pedestrian in a road traffic accident?
- Was this because of your eyes?
- Have you increased the power of your light bulbs?
- Have you fallen:
  in the house;
  on crossing the street;
  elsewhere outside?
- Have you changed the furniture or decoration of your house because of your eyes?
- Have you ever noticed that you have a blind spot?
- Have you ever seen double?
- When is your eyesight at its best and when is it at its worst?

Physical examination

The examination begins with inspection of the patient's spectacles, which may need cleaning or replacement. The pupils and the reflexes are checked, and care is taken not to miss an artificial eye. Ophthalmoscopy reveals the signs of cataract, glaucoma, optic atrophy, hypertensive and diabetic retinopathy, retinal detachment, and macular degeneration.

Functional examination

An initial impression of impaired vision is sometimes obtained when the doctor puts out a hand to shake the patient's, and the patient 'misses' the doctor's hand. The patient is asked to describe what the examiner is wearing, in order to obtain an initial idea of the level of visual function. He or she is then requested to read a newspaper with varying sizes of print, or the label on a medicine bottle. The effect of dull and bright lighting on performance is determined.

Focal vision is further tested by asking the patient what can be seen through the window or in the room, or by testing ability to put an electric plug into a socket or a straw through the hole in a soft-drink carton.

The initial testing of visual fields by confrontation is sometimes unsuccessful; and detailed mapping requires specialist help. An alternative which is well suited to the ward-round involves distributing four or five people over the visual scene at some distance from the patient. The doctor stands straight in front of the patient, ensures that the patient keeps his gaze fixed on him, and asks the patient whom he can see.

Ambient vision is tested by asking the patient to walk across a floor area containing low items of furniture, like coffee tables. A tendency to bump into these suggests impairment of ambient vision, but may also be due to unilateral visual agnosia or to the tunnel vision of advanced glaucoma.

Visual rehabilitation

Barriers

Patients whose visual disability is due to diabetic retinopathy may have accompanying peripheral neuropathy, which makes it difficult for them to use Braille or moon type, or to operate a long cane.

Elderly people may fail to achieve their full potential for making effective use of residual vision because of unavoidable reasons such as accompanying illness and/or impaired mobility or cognition. Too often the reasons are avoidable. They include fatalistic acceptance of the disability and low expectations of overcoming it; low availability and low awareness of rehabilitation services for the elderly; unawareness of the value and availability of visual aids; and inability to pay for aids and services.

Principles of rehabilitation

- Visual impairment is not an inevitable accompaniment of old age. Old people who cannot see well should expect to see better.
- The optician (optometrist) should be visited at least once every two years. Those who cannot afford the service or who cannot reach the optician should be identified and assisted.
- Low-wattage bulbs are a false economy. Vision is greatly improved by higher levels of illumination.
- Cluttered furniture limits mobility in the home and invites accidents. Spaces should be clearly defined.
- Walls should be white and furniture dark to offer maximum contrast.
- Large patterns and vertical stripes on walls and curtains provide a better visual milieu than do small and fussy patterns.
- Visual aids which are well selected to meet the needs of the patient should be prescribed.
- There is no shame in being visually handicapped and no harm in letting strangers know by the use of a white stick.

Methods of visual rehabilitation

The physician assesses to what extent the patient is capable of learning to regain independence that has been allowed to lapse because of visual impairment. Does the patient have the mobility, the dexterity, the sensitivity, and the mental flexibility to undergo the intensive training that is offered in a visual rehabilitation centre, assuming that there is such a facility in the locality? 'Poor motivation' is not a reason for witholding treatment. Poor motivation is often caused by exposure to the negative attitudes of others, and melts away when patients are shown what can be achieved.

Visual rehabilitation may include:

- mobility training, using an assistant, a long cane or rarely a dog (dogs are usually unsuitable for old people because they need a lot of care, and because they may outlive their master or mistress).
- domestic training, including food preparation and kitchen aids;
- rearranging the home, including provision of lighting, advice on furniture and decoration, domestic equipment;
- reading and viewing aids, and use of radio and tapes;
- learning typing, use of word processors and computers;
- learning Braille or moon type, which has been successful occasionally even in patients in their 90s.

Conclusion

When so much can be done to help victims of visual impairment it is unfortunate that many potential beneficiaries of care are overlooked. Visual impairment is rarely the main reason for an elderly person being referred to a geriatric unit, but is a frequent accompaniment of other diseases, and adds greatly to disability.

Notes and references

Age changes

Cullinan (1986) and Keeney and Keeney (1980) describe the common changes in structure and function of the eye in later life. There is impairment of the siphoning of tears leading to the 'rheumy' eye to be seen in Rembrandt's portraits. The lens darkens, so that a higher level of illumination is required in order to maintain the same level of visual function in subdued light. The pupil becomes smaller (Fledelius 1988). The anterior chamber becomes shallower from age 30 onwards, with corresponding thickening of the lens (Olbert 1988). Increased nuclear opalescence of the lens is physiological and does not betoken cataract. There is peripheral retinal atrophy and pigmentary degeneration. The cup/disc ratio remains less than 0.3.

Cataract

In the early stages, areas of discontinuity in the lens lead to scattering of light (Tripathi and Tripathi 1983). Breakdown of lens tissue and imbibing of fluid makes the lens more convex and improves the vision of myopics: 'If you can do without your spectacles you will need your lens out.' Posterior lens opacities, even when quite small, 'plug' the lens and diminish the amount of light reaching the retina.

The two main clinical variants are soft or cortical cataract, characterized by tightly packed bundles of swollen fibres; and hard or nuclear cataract, in which sclerosis occurs in the fibres of the lens core. In anterior and posterior subcapsular cataracts, necrotic changes occur in the subcapsular connective tissue, with degeneration, proliferation, and enlargement of cells.

Exposure to ultraviolet light is a significant risk factor for cortical cataract. Sun-glasses afford much less protection than the wearing of a hat with a brim (Taylor *et al.* 1988).

Glaucoma

The correlation between intraocular pressure and loss of visual field is less close than was formerly believed; while the effect of lowering intraocular pressure on preventing field loss may be less than was hoped (Vogel *et al.* 1990). Other risk factors include race, hypertension, and the previous occurrence of migraine (Wilson 1989). Angle-closure glaucoma occurs more commonly in hypermetropic patients and in women (Fledelius 1988).

Macular degeneration

Loss of the basement layer of the macular epithelium allows fluid, blood, and new vessels to pass out into the subretinal space (Davidorf 1981). The new vessels may bleed into the macular region. Round, irregular, yellow areas are seen deep to the retina in the macular area, and hyalinized nodules may appear. The width of the exudates exceeds that of the disc. There are clumps of pigment all over the retina; but there is no relation between the ophthalmoscopic appearance and the visual function. An early sign of macular degeneration is the loss of the little, dancing, prefoveal light reflex. If a central haemorrhage occurs it leaves a large central scotoma which may resolve after a year, leaving a large, brown patch. There is loss of central vision, with distortion, difficulty in reading, sewing, driving, etc., but not in getting about, because peripheral vision is spared.

Diabetic retinopathy

Podolsky and L'Esperance (1980) distinguish between 'background' and 'proliferative' retinopathy. In the background type, there are venous dilatation and microaneurysms, punctate haemorrhages, hard retinal exudate, and retinal oedema. Later there are dilated capillary-shunt vessels, cotton-wool spots, venous beading, and macular oedema. This causes reduced central visual acuity. Twenty years after the diagnosis of diabetes is made, 80 per cent of subjects are affected by this form of retinopathy.

In the proliferative variety, new vessels form on the disc and elsewhere. The condition is more common in smokers. Vitreous haemorrhage, retinal detachment, and total blindness may occur. Five years after the onset of the proliferative type, 60 per cent of subjects are blind.

Assessment

Deterioration of visual function occurs slowly in old people (Milne 1979, 1985). Visual testing using Snellen types shows little relationship to everyday visual functioning (Neumann *et al.* 1988). The test results are influenced by the level of illumination, the presence of direct sunlight, the number of letters per line, and reading and cognitive skills.

Among the more subtle aspects of visual function which undergo deterioration in later life are the ability to perceive the direction of self-motion (Warren *et al.* 1990).

Management

The average eyedrop measures from 40 to 75 μml, whereas the capacity of the average eyelid in old people is only 30 μml (Johnson 1986). Part of each eyedrop is spilled or washed away in the tears, but this is overcome by the use of Ocuset membranes.

Nearly one-third of mentally unimpaired patients in a continuing-care facility benefited from evaluation of their vision (Fenton *et al.* 1975). In most cases this was by adjusting, replacing, or providing spectacles.

References

Cullinan, T. (1986. *Visual disability in the elderly*. Croom Helm, Beckenham, Kent.

Davidorf, F. J. (1981). Retinal breakdown in the ageing eye: what are the consequences? *Geriatrics*, 36, 103–7.

Fenton, P. J., Arnold, R. C., and Wilkins, P. S. W. (1975). Evaluation of vision in slow stream wards. *Age and Ageing*, **4**, 43–8.

Fledelius, H. C. (1988). Refraction and eye size in the elderly. *Acta Ophthalmologica*, **66**, 241–8.

Johnson, D. H. (1986). New approaches to an old problem: the medical treatment of glaucoma. In *Year book of ophthalmology* (ed. J. T. Ernest and T. A. Deutsch). Year Book Medical Publishers, Chicago.

Keeney, A. H. and Keeney V. T. (1980). A guide to examining the ageing eye. *Geriatrics*, **35**, 81–91.

Milne, J. S. (1979). Longitudinal studies of vision in older people. *Age and Ageing*, **8**, 160–6.

Milne, J. S. (1985). *Clinical effects of ageing: a longitudinal study*. Croom Helm, Beckenham, Kent.

Neumann, A. C., McCarthy, G. R. Steedle, T. D., Sanders, D. R., Raanan, M. G. (1988). The relationship between indoor and outdoor Snellen visual acuity in cataract patients. *Journal of Cataract and Refraction Surgery*, **14**, 35–9.

Olbert, D. (1988). Relation of the depth of the anterior chamber to the lens thickness: clinical significance. *Ophthalmological Research*, **20**, 149–53.

Podolsky, S. and L'Esperance, F. A. (1980). Diabetic retinopathy: update on therapeutic advances. *Geriatrics*, **35**, 67–73.

Taylor, H. R. *et al.* (1988). Effect of ultraviolet radiation on cataract formation. *New England Journal of Medicine*, **319**, 1429–33.

Tripathi, R. C. and Tripathi B. J. (1983). Lens morphology, aging and cataract. *Journal of Gerontology*, **98**, 258–70.

Vogel, R, Crick R. P., Newson, R. B., Shipley, M., Blackmore, H. and Bulpitt, C. J. (1990). Association between intraocular pressure and loss of visual field in chronic simple glaucoma. *British Journal of Ophthalmology*, **74**, 3–8.

Wilson, R. P. (1989). Glaucoma care. In *Year book of ophthalmology*, (ed. J. T. Ernest and T. A. Deutsch, pp. 43–9. Year Book Medical Publishers, Chicago.

Aphorisms

Of sensory disability

- The deaf hear but the sound is distorted, like a badly tuned radio. The blind see but the picture is unclear, like a faulty television.
- To talk of 'admitting' to blindness or deafness implies the commission of an offence against society. The white stick and the hearing aid are statements of need, not badges of shame.
- Why is it so difficult to say, 'I do not hear well, please speak slowly and clearly', and 'I do not see well, please turn up the lights'?

Of visual impairment

- Whoever has driven with a frosty windscreen has experienced the anxiety of visual handicap.

- 'Blurred' is to vision as 'dizzy' is to balance; it reduces varied sensory experiences to a single word.
- The vision of a diabetic changes with changes in the blood sugar.
- You can tell a cataract patient by the peak on his cap.
- There are two kinds of vision, focal and ambient. Focal vision answers the question, 'What?' Ambient vision answers the question, 'Where?' Focal vision is conscious, central, and sharp; ambient vision is unconscious, peripheral, and blurred.
- Loss of focal vision impairs reading; loss of ambient vision impairs moving.

Of visual rehabilitation

- To improve the sight increase the light.
- Low vision aids improve focal vision; mobility training improves ambient vision.
- In the land of the blind the one-eyed man is king. But in the land of the seeing the one-eyed man may not get his low vision aid or mobility training.

17 *Auditory impairment*

Introduction

Two out of three people over the age of 70, and three out of four over the age of 80, experience difficulty in hearing and understanding speech. While only a few are unable to hear any speech at all, many find it difficult to comprehend conversation which is not addressed directly to them, or which involves more than two persons. Yet many old people with significant hearing impairment are unaware of their loss; and do not ask others to make allowance for their disability.

It is necessary to make sure how much an elderly patient hears and understands; otherwise communication is ineffective, and the impression is that the patient is unintelligent, uninterested, apathetic, or depressed.

The following material will be presented in this chapter, from a non-specialist view:

- the nature of speech comprehension;
- pathological changes affecting hearing;
- testing speech comprehension;
- management of hearing impairment.

Speech comprehension

Comprehending speech involves detecting the signal and decoding it. Speech comprehension is aided by speech music and by gesture. It is adversely affected by environmental sound.

Signal detection

The quality of the signal depends on the clarity of the speaker, and the state of the middle and inner ear. When one person speaks directly to another in a quite, non-reverberant room the signal is free of 'noise' and conditions are suitable for detecting speech. However, most conversations take place against a background of other noises, and the speech signals have to be differentiated from the 'noise'.

The speech signals consist of a rapid sequence of consonant and vowel sounds which vary in amplitude, duration, frequency, and significance. Compared to consonants, vowel sounds have a higher amplitude, a longer duration, and a lower frequency, but make less contribution to meaning. The string of letters:

EAI O I OO I AE IE

seems incomprehensible; but the string:

HRNG LSS S CMMN N LT LF

fairly clearly means 'Hearing loss is common in late life'. In this example of written speech the meaning of the vowels could not be guessed without the consonants, but the meaning of the consonants could be guessed without the vowels. Something analagous happens in auditory comprehension, where detecting the consonants is crucial to understanding.

Signal decoding

The signals detected by the ear undergo many stages of processing in the temporal cortex before they are decoded and comprehended. The time available depends on the rate of speaking. Before decoding can commence the speech signal is separated from the masking effect of background noise. It is also cleared of internal noise, due to the masking effect of loud, long-duration vowels on soft, short-duration consonants. The signal then proceeds to the stages of interpretation. The perceived sound is compared with an internal 'dictionary' of previously stored sounds, and an estimate is made of its meaning. This estimate employs information about the speaker, his or her accent, the context of the sentence, and the subject of conversation.

For example, the sound indicated by the spelling [MOOSS] might mean the animal 'moose' or the dish 'mousse' or the poet Robert Burns' way of addressing a mouse. In determining the meaning the listener takes into account whether the speaker is a Canadian trapper, a French chef, or a Scottish farmer; whether the conversation is about animals, food, or literature; and whether the preceding word is 'ferocious', 'strawberry', or 'wee'.

Long words are distinctive, and there is little ambiguity about their meaning. Examples are 'spectacles' and 'umbrella'. Short words like 'moose' are less distinctive, and more difficult to differentiate from similar sounding words.

A weakly heard signal increases uncertainty, and more time is required for processing. Speed is a critical determinant of intelligibility. The belief that foreigners speak fast is a misinterpretation. It only seems fast because of the need for more time to process the unfamiliar signal.

Speech music

Spoken language has rhythm, timbre, accent, and tone; and variations of speed, pitch, and volume. These are called the prosodic elements of speech, or speech music, and they aid and enrich the quality of communication. The world 'hello' can be said in ways which can convey delight, surprise, relief, curiosity, or indifference. The question 'Are you riding my bike?' can be given five different meanings by placing the emphasis on each word in turn. The expression and comprehension of speech music are impaired in people with hearing loss.

Gesture

Gesture, or body language, including the movement of the lips and the facial muscles, adds to the expression and comprehension of spoken language.

Summary

Processing speech therefore involves:

- hearing what is said;
- listing possible meanings;
- identifying the speaker;
- being familiar with his or her accent;
- understanding the subject of conversation;
- hearing the preceding words;
- retaining the sound in the memory;
- hearing the succeeding words.

Environmental factors

Comprehension of speech is affected by:

- the amount and nature of background noise;
- the degree of reverberation;
- the number of speakers;
- the noise level of their speech;
- the range of accents;
- the visibility of the speaker.

Comprehension is easiest when there is only one speaker sitting face to face with the listener, clearly visible to him or her, speaking slowly and distinctly, and using gesture and speech music. It is most difficult in a crowded reverberant room exposed to traffic and other noise, where many people are speaking at once, and where the listener has to change the focus of his or her interest frequently.

Comprehending speech comprises hearing, seeing, remembering, and understanding. It is a time-consuming process which involves communication between the temporal lobe and the parts of the brain which store and process visual, verbal, and cognitive information.

Pathological changes affecting hearing

Hearing in old age may be adversely affected by:

- childhood middle-ear disease;
- work-time noise exposure;
- other longstanding ear pathology;
- degenerative changes in the organ of Corti and the auditory nerve;
- loss of neurones in the auditory pathway;
- stiffening of the basal membrane;
- pathological changes in the temporal cortex;
- visual impairment.

Sensorineural hearing loss, due to changes in the inner ear, characteristically affects the upper range of speech frequencies (2–4 kHz), which correspond to consonants; and tends to spare the lower frequencies (1–2 kHz), which correspond to vowels. A feature of sensorineural loss is the phenomenon of *recruitment.* This is an overflow of loud sounds into adjacent frequencies, rather like what happens in a badly tuned radio. The resulting noise jars on the ear of the listener, and adds to the difficulty of interpreting the intended sound.

The speed and efficiency of signal processing in the temporal lobe is adversely affected by pathological changes in that part of the brain. These are of common occurrence in cerebrovascular disease, and especially in Alzheimer's disease, in which the temporal lobe is among the earliest parts of the brain to be affected. But even in the absence of dementia, many people show evidence of reduced filtration and impaired signal processing.

Those whose hearing is impaired cease after a time to hear their own pitch variations. This renders their speech flat and monotonous, and gives to the listener an impression of apathy or depression.

Tinnitus

Tinnitus is a common and troublesome accompaniment of hearing impairment. It may occur temporarily, either spontaneously or induced by noise, drugs, or toxaemia. Continuous tinnitus may arise from the cochlea, as a result of lifetime noise exposure or sensorineural deafness. Patients describe the sensation as being like blowing wind, the roar of waves, ringing of bells, crackling, beating, and whistling. It may or may not be synchronous with the pulse. The noise may be barely noticed; or it may dominate consciousness; and some patients are driven to suicidal thoughts because of it.

Testing speech comprehension

The doctor dealing with elderly patients should:

- determine the patient's ability to comprehend speech at the outset of the interview;
- make the conditions as favourable as possible for the patient to understand;
- seek the cause of the disability;
- check hearing aids;
- refer in suitable cases to a specialist.

Determining the patient's ability to comprehend

Clues to hearing difficulty are a puzzled look, a shy smile, an inclined head, a hand raised behind the ear, frequent nods, and tentative answers. An effective test is to determine the patient's ability to repeat a two-digit number like 'twenty-seven' whispered in each ear from a distance of 10 cm. Tuning-fork tests are not specially useful. Audiometry quantifies hearing function and assists diagnosis. Patients should be referred to specialists for this.

Making the conditions favourable

The following are guidelines for speaking to patients with hearing impairment:

- avoid background noise;
- sit close to the patient;
- do not raise the voice;
- speak slowly;
- enunciate distinctly;
- say a little at a time;
- repeat what has been said in different words;
- preserve speech music and gesture;
- make sure that the patient can see your mouth.

The common behaviour of talking loudly to deaf people is poignantly inappropriate. Loudness barely improves the intelligibility of consonants, but increases the masking effect of vowels and induces the jarring noise of recruitment.

The situation is summed up in a dialogue between doctor and deaf person.

- The doctor asks a question.
- The deaf person says, 'Don't mumble'.
- The doctor repeats the question in a louder voice.
- The deaf person says, 'Don't shout'.

'Don't mumble' means that the consonants cannot be distinguished. 'Don't shout' means that the loud vowel sounds are jarring.

Interviewing patients with hearing impairment

Once the existence of hearing impairment has been detected patients are asked:

- What have they noticed about their hearing?
- How has the disability progressed over time?
- Have they experienced tinnitus?
- What have been the social and psychological effects?
- What action have they taken?

Many patients, especially those who live alone, have little awareness of the insidious progression of hearing loss, and suffer little social deprivation. Others recognize that they are deaf but accept this as an inevitable consequence of ageing. A few are profoundly depressed.

The following questions help in evaluating the condition.

- Which of the following are difficult to hear:
  door bell;
  telephone bell;
  telephone conversation;
  voices in a crowded room;
  voices when more than one person is speaking at a time;
  children;
  women's voices;
  men's voices?
- Do you hear noises in your ears? Describe these and say when they come on.
- Which of the following activities have been abandoned because of hearing difficulties:
  parties;
  visiting friends;

visiting family;
going to the pub;
going to the cinema or theatre;
going to church;
crossing the road?

- Which of the following have you been accused of:
  turning the radio or television up too loud;
  shouting at others;
  mumbling;
  being stupid?
- Which of the following feelings have you experienced:
  feeling stupid;
  feeling lonely;
  feeling depressed;
  avoiding other people?

Seeking the cause

Wax in the ears is not often the sole cause of deafness, and soft wax incompletely filling the meatus probably makes little difference to hearing. Hard wax pressing on the drum and completely filling the meatus is a different matter, and its removal, which is not easy, may dramatically improve hearing.

Chronic suppurative otitis media was common when the present generation of old people were young, and chronically scarred and perforated ear drums are responsible for hearing loss. Other conditions require specialist assessment.

Patients may not mention tinnitus spontaneously, but it should always be asked about. It is resistant to most forms of treatment, although success is claimed with modern methods of suppression.

Hearing aids

Patients may be divided into:

- those who have no aids and do not need them;
- those who have no aids and need them;
- those who have aids and do not wear them;
- those who wear aids and do not use them properly;
- those who use aids properly.

The last group is the smallest.

Hearing aids comprise a power source, a receiver, a switch, a volume control, and an amplifier, with in some cases a filter. Body-worn devices are small. Behind-the-ear or within-the-ear devices are extremely small.

Obstacles to the correct use of a hearing aid

These are emotional, cognitive, manual, and auditory.

Emotional

People rightly or wrongly feel shame at being deaf. Deaf, old people appear as figures of mockery and derision in plays from Shakespeare's time or earlier right up to today's television sit-coms. Deafness generates impatience and intolerance in unsympathetic people.

It makes sense to announce one's deafness to the world and to encourage others to speak slowly and distinctly. But a hearing aid is perceived by many old people as a badge of shame and is emotionally rejected.

Cognitive

A hearing aid is complex in shape and in operation. It is difficult to fit the mould into the ear, and difficult to understand the controls. Hearing aids are unsuitable for people with brain failure, who have difficulty in learning the proper placement of the aid in the ear and the processing of the altered auditory signal.

Manual

The degree of dexterity necessary to place the aid in the ear, to operate the controls, and to change the battery may be beyond the skills of those with arthritis or sensory loss.

Auditory

Hearing aids make sounds louder but not necessarily clearer. They amplify vowels as well as consonants, and background noise as well as foreground noise. They do not necessarily improve discrimination. Mentally flexible users adapt to the altered sound that they provide and find them effective; but many users reject them after the initial disappointment and do not persevere.

Management

Patients with significant hearing loss and those with tinnitus are referred to specialists for full diagnosis and management. The following actions can be initiated by the geriatrician and his or her multidisciplinary team.

The patient should be encouraged not to conceal the deafness, but to commence every conversation by asking the person to whom he or she is speaking to speak slowly and distinctly, but not to raise their voice.

The ears should be kept free of wax. Unnecessary environmental noise should be reduced. The door and telephone bells should be made louder and of lower pitch. A flashing-light device should be activated when the door or telephone rings. A vibrating device can be placed on the pillow.

When a hearing aid is supplied there should be a loop induction coil so that radio and television can be heard with the sound turned down.

Hospitals should have portable communication aids, mechanical or electronic, to facilitate communication with the very deaf.

Conclusion

Auditory impairment impoverishes the sensory world of elderly people and deprives them of the sounds which enrich the daily life of the young. The impairment is too often denied or minimized by the victims. Their low expectations should not be shared by the medical profession. Every elderly patient who attends a geriatric service should undergo a simple assessment of auditory function, and should be referred for further treatment.

Notes and references

The subject was fully reviewed by Hinchcliffe (1983), and by Salomon (1986), while Markides (1972) deals with the rehabilitation of the adult deaf. Estimates of the prevalence of hearing loss in the community were provided by Gillhome Herbst and Humphrey (1980) and by Davis (1983). Bergman (1985) describes age changes in the comprehension of spoken language.

Pathology

The pathological changes which characterize presbyacusis include: atrophy of the organ of Corti and the portions of the auditory nerve in the basal end of the cochlea; atrophy of the stria vascularis; stiffening of the basal membrane; and loss of neurones in the auditory pathway and cochlea (Schuknecht 1964; Suga and Lindsay 1976).

Functional impairment

Sensorineural hearing loss is characterized by reduced sensitivity, increased susceptibility to masking of consonants by vowels, increased masking by background noise and by echoes in reverberant rooms, difficulty in determining changes of pitch and the 'spectral shape of speech sounds'; difficulty in following the spectral shape of speech sounds, and in following rapid changes in amplitude, frequency, and pitch (Summerfield 1987).

Hearing loss is associated with advanced age, impaired cognitive function, and reduced quality of life, but only weakly with depression (Gillhome Herbst and Humphrey 1980; Peters *et al.* 1988; Bess *et al.* 1989). Walsh and Eldredge (1989) found a high level of isolation, frustration, powerlessness, and resentment of discrimination among aged people who had been deaf all their lives.

Assessment

Many patients with impaired hearing are unaware of this (Milne 1976); but testing with the whispered voice is effective in detecting significant hearing loss (Macphee *et al.* 1986).

Hearing aids

The types of hearing aid and their use were described by Lorrado (1988). Use declined with age; and six months after being supplied with a hearing aid, only one-third of elderly patients were using them often or all of the time (Stephens 1977; Littlejohns and John 1987).

Summerfield (1987) recommended the use of aids in both ears; directional microphones to attenuate echoes; and filtering the speech, with differential amplification of high-frequency sounds. However, Chung and Stephens (1986) believe that adjustment to binaural aids is even more daunting than to single aids, and is resisted even by younger patients.

Tinnitus

Coles and Hallam (1987) suggested the use of hearing aids, masking, and counselling. Al-Jassim (1987) claimed excellent results in young people by using a Walkman (portable) stereo system as a masking agent.

References

Al-Jassim, A. H. H. (1987). The use of the Walkman mini-stereo system in the management of tinnitus. *Journal of Laryngology and Otology*, 101, 663–5.

Bergman, M. (1985). Age-related changes in the comprehension of spoken language. In *Recent advances in geriatric medicine* (ed. B. Isaacs), pp. 141–56. Churchill Livingstone, Edinburgh.

Bess, F. H., Lichtenstein, M. J., Logan, S. A., Burger, M. C. and Nelson, E. (1989). Hearing impairment as a determinant of function in the elderly. *Journal of the American Geriatrics Society*, 37, 123–8.

Chung, S. M. and Stephens, S. D. G. (1986). Factors influencing binaural hearing aid use. *British Journal of Audiology*, **20**, 129–40.
Davis, A. (1983). The epidemiology of hearing disorders. In *Hearing and balance in the elderly* (ed. R. Hinchcliffe). Churchill Livingstone, Edinburgh.
Gillhome Herbst, K. and Humphrey, C. (1980). Hearing impairment and mental state in the elderly living at home. *British Medical Journal*, **281**, 903–5.
Gillhome Herbst, K. (1983). Psychosocial consequences of disorders of hearing in the elderly. In *Hearing and balance in the elderly* (ed. R. Hinchcliffe). Churchill Livingstone, Edinburgh.
Littlejohns, P. and John, A. C. (1987). Auditory rehabilitation: should we listen to the patient? *British Medical Journal*, **294**, 1063–4.
Lorrado, O. J. (1988). Hearing aids. *British Medical Journal*, **296**, 33–5.
Macphee, G. J. A., Crowther, J. A., and McAlpine, C. H. (1986). A simple screening test for hearing impairment in elderly patients. *Age and Ageing*, **17**, 347–51.
Markides, A. (1977). Rehabilitation of people with acquired deafness in adulthood. *British Journal of Audiology*, Suppl. 1.
Milne, J. S. (1976) Hearing loss related to some signs and symptoms in older people. *British Journal of Audiology*, **10**, 65–73.
Peters, C. A., Potter, J. F., and Scholer, S. G. (1988). Hearing impairment as a predictor of cognitive decline in dementia. *Journal of the American Geriatrics Society*, **36**, 981–6.
Salomon, G. (1986). Hearing problems in the elderly. *Danish Medical Bulletin*, Suppl. 3.
Schuknecht, H. F. (1964). Further observations on the pathology of presbyacusis. *Archives of Otolaryngology*, **80**, 369–82.
Stephens, S. D. G. (1977). Hearing aid use by adults: a survey of surveys. *Clinical Otolaryngology*, **2**, 385–402.
Suga, F. and Lindsay, J. R. (1976). Histopathological observations of presbyacusis. *Annals of Otolaryngology*, **85**, 169–84.
Summerfield, Q. (1987). Speech perception in normal and impaired hearing. *British Medical Bulletin*, **43**, 909–25.
Walsh, C. and Eldredge, N. (1989). When deaf people become elderly: counteracting a lifetime of difficulties. *Journal of Gerontological Nursing*, **15**, 27–31.

Aphorisms

Of auditory impairment

- Deafness is recognized by the turn of the head, the nod of agreement, the vapid smile, the excess of 'Yes's', the raised volume on the television.
- Foreclosing a conversation which cannot be heard is like foreclosing a mortgage which cannot be paid. The deaf are evicted from the social world as the debtor is evicted from his home.
- The patient who says, 'I can't hear you doctor, I need my spectacles', teaches the role of vision in hearing.

- The rules for communicating with the deaf are, don't mumble, don't shout, be seen.
- Remove thy hair from thy lips and thy hand from before thy mouth. Then shall the deaf hear and rejoice.
- Helping a deaf person across a sentence is like helping a blind person across a street.
- To simulate deafness, tune the radio off the station.

Of tinnitus

- Tinnitus is an uncongenial companion who talks incessantly to one who does not wish to hear.
- The roar of the waves of the sea, the whistle of the wind through the trees, the rush of a great steam train, the beat of a mighty engine; these are the sounds which torment the victims of tinnitus.

Of hearing aids

- A miniature hearing aid is the triumph of vanity over utility.
- Miniature hearing aids have miniature controls. Deaf people have clumsy fingers. Hearing aids live in drawers. Find the connection.
- A cheap hearing aid is like a cheap newspaper; it magnifies and distorts, but does not discriminate.
- The whistling of a hearing aid is like the siren of a police car. Help is needed urgently.

18 *Depression*

Introduction

Illness in old age, with its attendant pain, discomfort, and loss of mobility and independence, seems from the perspective of youth a profoundly unhappy experience. But it is remarkable how many elderly hospital patients bear their misfortunes with fortitude, patience, and stoicism; how many remain good humoured and capable of appreciating and enjoying the few remaining pleasures that life offers; and how many hold on to the last vestiges of independence with determination and vigour.

There are others who find their lot harder to bear, who are ready to give up the struggle to remain independent. This is expressed in a variety of ways, from weeping to withdrawal, from complaining to demanding.

Among the physically ill patients who come under the care of geriatricians are many who are depressed and some who suffer from a depressive illness. The distinction is not easily made, but it is valid and has implications for management. The present discussion of depressive illness in elderly people is limited to the problems which the geriatrician with no special training in psychiatry encounters during the management of physically ill patients.

Interviewing the patient

The geriatrician has to determine:

- Is the patient ill?
- Is he depressed?
- Is he depressed because he is ill?
- Is he ill because he is depressed?
- What factors other than illness contribute to his or her depression?

The stages of the interview are:

- initial impression;
- history of current physical illness;
- previous physical illness;
- previous psychiatric illness;
- history of losses;
- emotional state;
- attitude to death.

An *initial impression* of the patient's mood is gained from facial expression, tone of voice, and readiness to provide information. The patient may exhibit sadness, impatience, resentment, or withdrawal. He or she may spontaneously express delusions of unworthiness, of guilt, or of changes in his or her own body. Alternatively the delusions may be of paranoid content, or there may be visual or auditory hallucinations. Failure to answer simple questions may be due to accompanying brain failure or deafness. A rule of thumb for distinguishing depression from brain failure is that the depressed patient does not answer, while the patient with brain failure gives the wrong answer.

The history of the *current physical illness* follows the usual lines. Physical diseases which are frequently accompanied by depression include stroke, aphasia, Parkinson's disease, hepatic failure, hypothyroidism, anaemia, infections, amputation, rheumatoid arthritis, deafness, and malignancy. Drugs which induce depressive symptoms include antihypertensive and antirheumatic substances.

Depression may be associated with isolated physical symptoms for which no physical cause is apparent, such as facial pain and abdominal pain; or it can be manifest as hypochondriasis with multiple complaints of pain, pruritus, constipation, and strange sensations, which are experienced all over the body.

The questioner determines the effects of the current illness on the patient's daily life, and the presence or absence of the so-called biological features of depression, especially disturbances of appetite, bowels, and sleep pattern, and the attitudes to these. Questions are asked about the consumption of drugs and alcohol in the past and more recently.

The history of *previous physical illness* distinguishes between the patient who has 'never had a day's illness in his life' and the one whose body has grown inured to the restrictions which illness imposes on it.

The personal and family history of *previous psychiatric illness*, including depression, is explored. In the past, depressive illness may have been labelled 'nervous breakdown' and may have led to a prolonged stay in a mental hospital, or to a period of unemployment. The next stage is to determine what *losses* the patient has suffered, and how he or she has reacted to these losses. The list includes:

- physical health;
- mobility;
- independence;
- work;
- spouse;
- other family members and friends;
- income;
- home, garden, motor car;

- participation in sport or social activity;
- faith, or formal religious affiliation.

There may be many more. Old age, however, is not only about loss, and the patient should be asked also about gains ... new interests, new friends.

The questioning then moves on to exploring the patient's *emotional state*. Depression rating scales are sometimes used for this purpose. A less formal approach is to begin by finding out from the patient what an average day at home is like.

- What does the patient do that gives him or her pleasure and what causes grief or pain?
- Does he look forward each morning to the coming day, or does he dread it?
- Whom does he see each day?
- Does he enjoy seeing them?
- Does he write or receive letters?
- Does he use the telephone?
- Does he feel lonely?
- Does he speak to himself, and if so what does he say?
- Does he speak to God?
- Is there anyone to whom he confides his or her thoughts?
- Does he think a lot about people who are dead and gone?
- Does he have regrets for the past?
- How does he view the future?
- Has he financial concerns?
- Does he ever burst into tears for no reason?
- What comes into his or her mind as he lies awake in the morning before rising:
- What frightens him or her?
- How does he react to fear?
- How friendly or unfriendly are other people?

The questioning is then directed towards *death*. Some doctors are hesitant to introduce this topic, fearing that it will distress the patient. Rarely is this the case. Most elderly people welcome the opportunity to talk about a subject that is very much on their mind. They are close to and familiar with death, and for many it is the next experience that they have to look forward to. A doctor who hesitates to introduce the subject with an elderly person should consider whether the concern is to avoid being embarrassed him- or herself. S/he should be reassured that elderly people are very matter of fact about death. If they do not wish to talk about the subject they say so.

The doctor proceeds with circumspection, reviewing each reply before deciding whether to venture further. The exploration should not be abandoned if the patient begins to weep. On the contrary this is often a desirable outcome; and it is even better when the patient confesses, 'Doctor, I have never told this to anyone before.' This is the sign that the doctor has, as it were, crossed to the other bank, reached the concealed part of the patient's world and gained his or her confidence.

Some doctors hesitate to ask patients about whether they harbour suicidal thoughts 'in case this puts the idea into their mind'. If the idea is not there, a question from a doctor is unlikely to put it there. It is far more worrying to think of the idea being there, dominating the patient's thoughts, and no one is prepared to talk about it. The idea is more likely to be acted upon if it is never given verbal expression. Asking about suicide can be life saving.

Following is a list of questions from which doctors may choose what seems appropriate at the time.

- Do you ever think about death?
- What do you think about your own death?
- Do you ever go to bed at night and hope that you will not wake up again next morning?
- Have you ever contemplated bringing your life to an end?
- Why have you done nothing about it?
- Do you believe in life after death?
- Do you think you will ever see your loved ones again?
- Are you afraid of death?
- Do you think that you will be punished for your sins?
- Do you think that you are already being punished for them?

Wishing to die and being depressed are not the same thing. Many a cheerful 95-year-old woman will say in the most matter of fact tone that she has lived quite long enough, more than she had reckoned with, and will be very happy to die when God calls her. She has outlived her friends and relatives, has done all she wanted to do, does not wish to become a burden to anyone else, and would now like to die, but does not propose to do anything about it. There are others who say that they pray every night for God to take them away, but who rise next morning ready to meet the challenges of the day. Others again say that they never think about death and do not want to discuss the subject; and there are also those who say that they want to go on living to the full until they die. What matters is not the wish for death, but the way in which that wish is expressed.

Interviewing relatives

Relatives are asked about the changes which they have observed in the patient's behaviour, appearance, interests, eating, drinking, and smoking.

- Do they enjoy talking to her or him or do they find them depressing?
- Has he a history of previous mental illness?
- Can they throw light on this previous mental state?
- How did he deal in the past with bereavement, illness, crisis?
- Was the marriage happy?
- Was his career satisfying and fulfilling?
- How did he relate to his own and other people's children?
- How did he react to being ill?
- Has he become demanding?
- Does he complain of neglect, loneliness, unhappiness?
- Does he demand attention?
- Is he appreciative?
- Does he behave like a spoilt child?
- Does he have many physical complaints?

Examination

The conventional physical examination contributes to the diagnosis by determining the degree of physical illness and the appropriateness of the patient's response. A patient may have all the features of 'classical' biological depression; but the discovery of a malignant tumour in the stomach or lung may fully or partially account for the symptoms. A patient may appear to have a hypochondriacal preoccupation with the bowels, but a rectal examination discloses faecal impaction, treatment of which abolishes the 'hypochondriasis'. Information is gathered by nursing and other staff in the following areas:

- attitudes to getting up in the morning, to dressing, washing, bathing, personal appearance;
- appetite, and amount eaten at and between meals;
- willingness to drink;
- interaction with other patients;
- interaction with visitors;
- enthusiasm for rehabilitation;
- frequency of complaints and demands on staff;
- shouting;
- attention seeking.

Food refusal, sometimes mistakenly called 'anorexia nervosa of the elderly' is often observed in depressed patients. It may be a sort of suicide, the only way open to the very weak and despairing to bring their suffering to an end. Sitting in a chair with the head bowed so that only the back of the head is visible, and lying in bed with the covers pulled over the head, are strongly suggestive of a depressed and withdrawn state.

Shouting for no apparent reason upsets relatives and carers, and is a frequent cause of complaint from other patients and their relatives when it occurs in an open hospital ward. This symptom is usually associated with advanced brain failure, but may also be a manifestation of depression, and is sometimes improved by antidepressant therapy.

Evaluation

This information provides insight into the patient's feelings, offers a channel of communication with the patient, and helps to place him or her in one of the following categories:

- not sad;
- sad but not depressed;
- depressed but not suffering from depression;
- suffering from depression;
- suffering from other psychiatric illness.

It is admittedly difficult to define these terms; but they help to identify which patients require referral to a psychiatrist, and in which the decision should be made to commence treatment for depression. This decision is not made lightly, because even the best antidepressant drugs have prominent side-effects; and electroconvulsive therapy, while it is often safe and effective, is a resource only for the specialist.

Management

In favourable cases, such as hypothyroidism and addisonian anaemia, treatment of the physical illness may dramatically reverse the depression. In conditions which require rehabilitation, such as stroke and amputation, depression bars the patient's cooperation in treatment, prolongs the disability, and may render the depression permanent. In these cases it is worth giving a trial of antidepressant treatment at an early stage. Patients who express suicidal wishes and those who refuse food need urgent referral to a psychiatrist and may require electroconvulsive therapy.

Notes and references

The nature of depression

Diagnostic criteria for depression were recommended by the American Psychiatric Association (1980). Murphy (1986) maintained that depression in old age is the same as at any other age; that its manifestations are independent of age of onset; and that it is possible, although not always easy, to identify endogenous and neurotic types. Others disagree. Pitt (1986) applied the name 'dysphoria' to patients with persistent states of sadness which did not fulfil the diagnostic criteria of depressive illness. Schwartz and Blazer (1986) distinguished between 'depressive symptoms', which are common in late life, and 'depressive illness', which is uncommon; while Breckenridge *et al.* (1986) made a similar distinction in relation to bereaved, elderly people. Abramson *et al.* (1978) defined 'learned helplessness' as a condition which lay on the borderland of depression, but which did not proceed to frank depression. Sweer *et al.* (1988) found a high prevalence of physical illness among hospitalized, depressive patients. Murphy (1982) identified social factors as contributing to depression, the most significant being lack of a confidant.

Pseudodementia

Dementia and depression may coexist; dementia may lead to depression; and some cases of depression may simulate dementia. There is, however, no agreement that there is a specific condition of 'pseudodementia' (Patterson 1986; Pitt 1986).

Rating scales

Hamilton's scale, one of the earliest, remains in wide usage in the United Kingdom (Hamilton 1960). In the United States, preference is given to the Geriatric depression rating scale (Yesavage *et al.* 1983); but this has low diagnostic accuracy in a nursing home population (Kafonek *et al.* 1989).

Suicide

Suicide is most common in late life (Sainsbury 1962), although not in all social, cultural, and cohort groups (Manton *et al.* 1987). Predisposing factors include isolation, recent bereavement, poor health, persistent pain, and clinically apparent depression (Catell 1988).

References

Abramson, L., Seligman, M., and Teasdale, J. (1978). Learned helplessness in humans: critique and reformulation. *Journal of Abnormal Psychology*, 87, 69–74.

American Psychiatric Association (1980). *Diagnostic and statistical manual of mental disorders*, (3rd edn). APA, Washington DC.

Breckenridge, J. N., Gallagher, D., Thompson, L. W., and Peterson, J. (1986).

Characteristic depressive symptoms of bereaved elders. *Journal of Gerontology*, **41**, 163–8.
Catell, H. R. (1988). Elderly suicide in London: an analysis of coroners' inquests. *International Journal of Geriatric Psychiatry*, **3**, 251–61.
Hamilton, M. (1960). A rating scale for depression. *Journal of Neurology, Neurosurgery and Psychiatry*, **23**, 56–62.
Kafonek, S., Ettinger, W. H., Roca, R., Kittner, S., Taylor, N., and German, P. S. (1989). Instruments for screening for depression and dementia in a long term care facility. *Journal of the American Geriatrics Society*, **37**, 29–34.
Manton, K. G., Blazer, D. G., and Woodbury, M. A. (1987). Suicide in middle age and later life; sex and race specific life tables and cohort analyses. *Journal of Gerontology*, **42**, 219–27.
Murphy, E. (1982). Social origins of depression in old age. *British Journal of Psychiatry*, **141**, 135–42.
Murphy, E. (1986). The concept of affective disorders in the elderly. In *Affective disorders of the elderly* (ed. E. Murphy), pp. 1–12. Churchill Livingstone, Edinburgh.
Patterson, C. (1986). The diagnosis and differential diagnosis of dementia and pseudodementia in the elderly. *Canadian Family Physician*, **32**, 2602–10.
Pitt, B. (1986). Characteristics of depression in the elderly. In *Affective disorders of the elderly* (ed. E. Murphy), pp. 40–52. Churchill Livingstone, Edinburgh.
Sainsbury, P. (162). Suicide in later life. *Gerontologia Clinica*, **4**, 161–70.
Schwartz, M. S. and Blazer, D. G. (1986). The distribution of affective disorders in old age. In *Affective disorders of the elderly* (ed. E. Murphy), pp. 13–39. Churchill Livingstone, Edinburgh.
Sweer, L., Martin, D. C., Ladd, R. A., Miller, J. K., and Karpf, M. (1988). The medical evaluation of elderly patients with major depression. *Journal of Gerontology*, **43**, M53–8.
Yesavage, J. A. *et al.* (1983). Development and validation of a geriatric depression screening scale: a preliminary report. *Journal of Psychiatric Research*, **17**, 37–49.

Aphorisms

Of depression

- More remarkable than the high prevalence of depression in old age is the low prevalence of depression in old age.
- Young doctors look on depression as the normal state of old people; older doctors look on acceptance as the normal state of old people.
- The art of diagnosing depression is to distinguish between being depressed and having plenty to be depressed about.
- The Ghosts of Sorrows Past return to haunt the aged.
- The fear of death in old age is less than the fear of life in old age.
- Depression is that disease which responds to antidepressant therapy. Resistant depression is that disease which does not respond to antidepressant therapy.
- Physical illness is the commonest cause of depression in late life.

Of the clinical features of depression

- A weeping adult has a tale to tell.
- As diabetes is diagnosed by a drop of urine, so depression is diagnosed by a drop of tears.
- You can diagnose depression from the back of the patient's head, if that is all you see of him when he sits.
- An anxious patient who knows what he is anxious about is not suffering from depression. An anxious patient who does not know what he is anxious about is suffering from depression.
- A patient who describes daily variation in the shape, size, texture, and colour of his motions may not be suffering from depression, but the doctor's thoughts tend to turn in that direction.
- Man says, I am sad; woman says, I am sore.
- The pain of depression never shifts, never lifts, never varies.
- The hypochondriac takes your hand to show you where the pain is.
- If you feel depressed after you have interviewed the patient he probably has depression.

Of the death wish

- The suicidal wish to kill themselves; the ill wish to die; the old wish to be dead.
- Those who wish to kill themselves need a psychiatrist. Those who wish to die need a physician. Those who wish to be dead teach us about old age.
- The wish to die expresses unfulfilment; the wish to be dead expresses fulfilment.

Of bereavement

- Most widows were married to the best husband in the world.
- Sanctification of the deceased is poor adjustment to his death.
- Loss of a spouse does not cause depression: it is the way of the world. Loss of a child causes depression: it is against the way of the world.
- Widows love their homes, their neighbours, their pets. Separation from these may initiate depression.
- It is is normal, common, and healthy for the bereaved to see the deceased, hear the deceased, speak to the deceased; and they should be told so.

Index